Table of Contents

Preface
About the Author
How to use this book
ABSTRACT

Chapter 1

Overview of a Programmable Logic Controller (PLC)
How the Ladder Logic program is supposed to function?
Speedometer electronic circuitry
System hardware
Block diagram of the PLC system connected to HMI, main PCB etc
Schematic diagram of the main control PCB
Schematic diagram of 2×7-segment digits display PCB

Chapter 2

Developing main control software for S7-300 PLC
Coding for the second intersection
Coding for the third intersection
The Emergency mode
Controlling the countdown timer displays
Addressing the countdown timer displays
Coding to display number 00 when system is on its emergency mode
Using System Function Blocks to display current Time and Date
Measuring speed of a passing car and comparing it with 30 Km/h speed limit

Chapter 3 -Adding an HMI display to our traffic light system project

Phase 18- Installation of a typical HMI LCD touch control panel
Phase 19- Altering the control program to establish connection to HMI
Screen design of HMI device

Chapter 4- Application of the WinCC Explorer as an HMI task

Phase 19- Developing an HMI application with SIMATIC WinCC
Questions and Answers related to this project
Conclusion
My product list

The project is about development of a practical traffic light control system using SIMATIC S7-300 as the main controller of the project. To develop the project, there are two objectives that must be accomplished which are:
1. Development of a traffic control system controlled by a PLC. In that case, a SIMATIC S7-300 PLC will be used to control the operation of the traffic light system in 4 intersections.

2. Programming with STEP7: STEP 7 software is used to program the PLC. There are three programming languages in the STEP7: In this project, most sections of the control program is developed using LDA programming language.

3. Application of SIMANIC WinCC and PM Designer HMI software programs to establish connection of PC monitor and 7 inch displays to make the usage of the system much more user friendly.

The purpose behind of developing a manual like this is an attempt to make learning Programmable Logic Controllers fun! Making connection with a familiar control system such as one full functional controller a PLC brings out the how and why latches, timers, limit switches, sensors, analog instructions, and relays are used. The programs in this manual were written for SIEMENS S7-300 controller using STEP 7 software but the ideas and program layouts are universal to any PLC.

Seyedreza Fattahzadeh holds a Bachelor of Science in Electrical Engineering degree from the University of Texas at Arlington. His major areas of expertise are digital systems, electronics, and computer engineering.

Seyedreza has extensive experience in the filed of programmable controllers. He worked for few industrial manufacturers as a programmable controller consulting firm. His industrial experience includes designing and implementing mid class automation systems, based on PLC.

Seyedreza has authored several other books about programmable controllers. You can find more info on these at the end of this book

This book is about programming a S7-300 PLC to function as traffic light controller. In my other books, I have discussed the basics of Programming with different models of PLCs. This book has been prepared for those who are already familiar with basic PLC instructions and now wish to challenge their knowledge by writing more complex industrial PLC programs: in this case, preparing a control program for a PLC to control a 4 intersections traffic light and related timing counters.

The hardware and software of the project is provided for you. Read the project specification carefully. Your job here is to re-write the programs. Use pen and paper, or use your PLC simulator or your own PLC (any brand!) to write the program. When you have done it, run your program and test its performance with any kind of software simulator you have. If your solution runs according to the problem software specification, it is fine, otherwise, use my solution against yours. Fortunately, STEP7 comes with an excellent simulator. Use it to check out how my solution works. And finally if you require, compare mine with yours to resolve and correct your own control program problems in the way that best works for you. By doing all these activities, at the end, you will learn and that is the goal of this project.

The program solution is done based on Ladder Logic language which is mostly used one in PLC application field nowadays.

If you wish, you can build your own controller hardware similar to mine or even tailor it to the one with extra features you like. By simulating your own program and checking its performance, you can improve your programming skills! And if you are stuck at any step, you can check out my solution against yours!

In any case, have fun and drop me a line about your progress on writing your PLC software projects. My email address is: srfattah@yahoo.com

Implementing a TRAFFIC LIGHT CONTROLLER using **a PLC** is the title of this project. This project is divided into two parts which are hardware and software. The hardware part for this project is a model of four way junction traffic light. The Control program is developed for a main street (S/N) with 4 intersections (4 E/W streets) to control their traffic lights. On locations where each secondary East-West road intersects the main one, I have two traffic light boxes each with 3 light indicators and a 1×2 digits countdown meter. **Thus, the purpose of the project is to control all these timers and traffic lights all in harmony**. The purpose of using a countdown meter is to remind drivers and pedestrians of the waiting time through counting down numbers to effectively reduce the rate of traffic accidents.

<u>Overview of a Programmable Logic Controller (PLC)</u>

A programmable logic controller (PLC) is an industrial computer used to control and automate complex systems. PLCs are a relatively recent development in process control technology. It is designed for use in an industrial environment, which uses a programmable memory for the integral storage of user-oriented instructions for implementing specific functions such as logic, sequencing, timing, counting, and arithmetic to control through digital or analog inputs, outputs, and various types of machines or processes.

A PLC is used to monitor input signals from a variety of input points which report events and conditions occurring in a controlled process. PLCs are used to control robots, assembly lines and various other applications that require a large amount of data monitoring and control.

PLCs are normally constructed in modular fashion to allow them to be easily reconfigured to meet the demands of the particular process being controlled. The processor and I/O circuitry are normally constructed as separate modules that may be inserted in a chassis and connected together through a common backplane using permanent or releasable electrical connectors.

Project objectives

To develop the project, there are two main objectives that must be accomplished. They are:
1- Constructing a model of a 4 × four way junction of a traffic light model.
2- Developing a ladder logic control program for the system.

Combining the software part and the hardware part to simulate a traffic light system

In this project we will develop a practical traffic light control system. It is primarily designed to be used as a training project to teach students how to write a complex control program through using network consisting of many timers, counters, basic digital, and special functional blocks, in total, about 180 networks.

The hardware of the project consists of the following functions:

1- It has eight (two digits each) LED countdown timers to remind drivers and pedestrians of the waiting time through counting down numbers to effectively reduce the rate of traffic accidents.

2- It has one (two digits) LED numerical display to show the calculated speed of the car passing between two IR sensors installed at the distance of 100 meters (the distance is defined in the control program). Based on the time it takes a car to pass these two IR sensors, PLC calculates the speed of the car and displays it on the 9th two digit numerical display in Km/h. Red LED is turned on if speed is > 30 Km/h, and the green LED is turned on if speed is < 30 Km/h.

3- It has one (six digits) LED numerical display to show the current time and date.

4- It has one control main PCB to feed data to numerical displays, and indicators (red, yellow and green LEDs), transfer depression of pushbuttons, and IR sensors to PLC via its Input or Output terminals.

How the Ladder Logic program is supposed to function?

We wish to have a control program, that when RUN, command is issued to the PLC, the output performance of the control program to function as follows:

1- We would like all stop lights at each intersection to turn on and off based on the data shown on table at figure 1.1. All **times** are in seconds. And figure 1.2, displays the location of each stop light and the LED countdown timers.

S/N	E/W	S/N	E/W
G1= 80	R2 = 85	G3 = 60	R4 = 65
Y1= 5	Y2 = 5	Y3 = 5	Y4 = 5
R1= 60	G2 = 55	R3= 50	G4 = 45
145	145	115	115

S/N	E/W	S/N	E/W
G5= 90	R6 = 95	G7= 85	R8= 90
Y5 = 5	Y6 = 5	Y7= 5	Y8= 5
R5 = 50	G6 = 45	R7= 50	G8= 45
145	145	140	140

Figure 1.1

From figure 1.1, notice that it takes about 145 seconds for the signal lights to complete their cycle at the (S/N) first intersection. After G1 turns on, and the counter starts counting down from 80 to 0 seconds, the yellow traffic light = 5, and finally, the red signal light turns on for 60 seconds. **Cycle time = 145 seconds**.

Driving within speed limit, it takes 600 seconds to get to the next intersection traffic lights. We have 10 min. × 60 = 600 seconds. Since the cycle time of second traffic lights is 115 seconds, and we want when the same car gets to the second traffic lights to receive a green signal light, then **25** seconds after G1 turns on, we want G3 to be turned on. Therefore, we can write 600 = (5 × 115 + 25 = 575 + 25). The third traffic lights (G5) have a cycle time of 145 seconds and it takes 30 minutes × 60 = 1800 seconds = (12 ×145 + 60 = 1740 + 60) to get to the third **signal light (from the second traffic lights)**. And we want the same car to get a green traffic light when it reaches the third one. Then, 60 seconds after G3 is on, G5 can be turned on. So far, we have 3 green signal lights turned on. Cycle time of G7 is 140 seconds, but it takes 40 minutes × 60 = 2400 seconds for the car to get from G5 to G7. The cycle time of G7 = 140 seconds. So we have 2400 = 40 minutes × 60 = 2400 seconds = (17 ×140 +20 = 2380 + 20) this means 20 seconds, after G5 turns on, G7 must also be turned on.

Now we have 8 traffic lights alternating the right of way granted to road users by displaying lights of a standard color (red, yellow, and green).

Figure 1.1A displays the initialization phase of the system (when system is turned on for the first time). It must be understood that, all **E/W** street traffic lights must be turned on after related **S/N** green light is turned on.

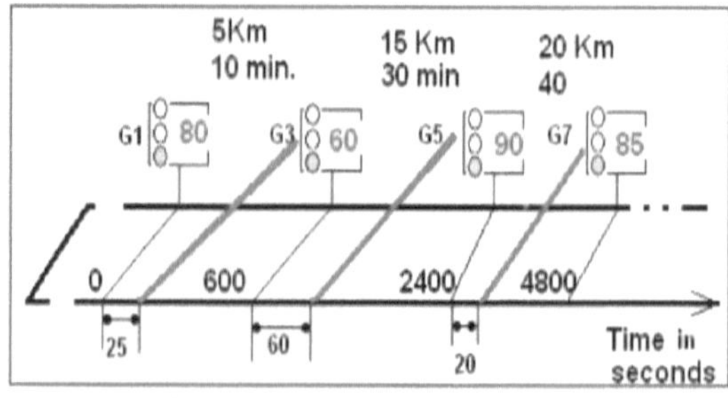

Figure 1.1A

1- Figure 1.3 shows what is the approximate distance between each intersection stop light. From figure 1.3, notice that the distance between first and second intersection is about 5 Km. Therefore, if the speed limit at that road is about 30 km/h, then from speed equation 5 km = 30 km/h × h or h = 1/12 × 60 = 10 minutes. Figure 1.2a displays the electronic schematic of S1 and S2 IR sensors. Figure 1.2b displays the physical view of each PCB designed to be installed for S1 or S2.

2- The system takes the current time and date from PLC's internal clock and shows it on a 6 digit seven segment numerical display. See figure 1.4. When the system is displaying the current time, 4 LEDs blink and 2 LEDs blink when it is displaying the current date.

3- At 24:00 PM, system goes into its Emergency mode, where all yellow LEDs on the S/N main road start blinking in yellow and the E/W lights in red. System returns to its normal operation at 7:00 AM when the lights turn on/off based on the regular schedule.

4- By depressing the **Emergency** pushbutton, we can also force the system into **Emergency** mode. Pressing the **STOP** and **START** pushbuttons in sequence causes the system to return to its normal operation. In Emergency mode, number 00 will show on all LED displays.

Speedometer electronic circuitry

Figure 1.2A displays the schematic circuitry of the IR sensor (Break-Beam Sensor). Notice that when circuit in figure 1.2 is energized, the **NO** terminal of the circuit is raised to 24 VDC. And when the IR ray is broken by presence of a car crossing S1 or S2, the relay is activated (NC = 1 or 24 VDC) and that is the time which PLC is notified that a car is crossing either sensor Figure 1.2B displays the physical view of the **IR sensor**. Electronic components shown in figure 1.2A are soldered to a printed circuit board to create the **IR sensor PCB** shown in figure 1.2B.

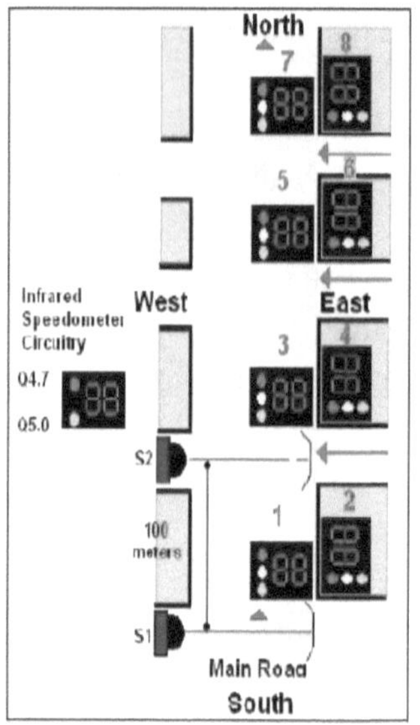

Figure 1.2

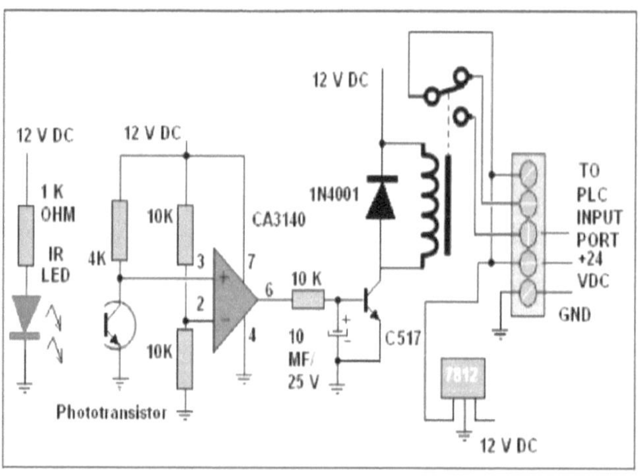

Figure 1.2A

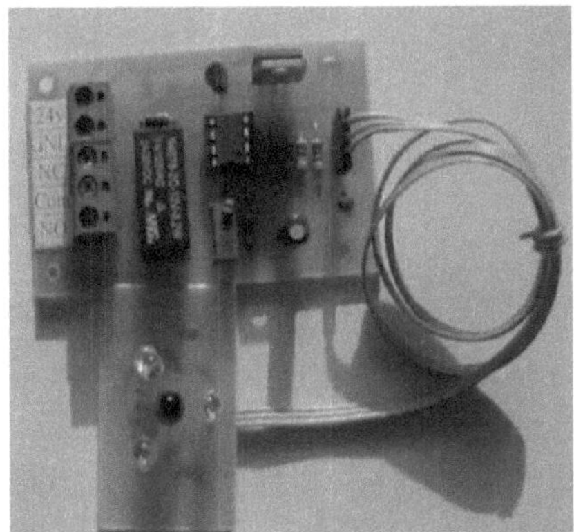

Figure 1.2B displays the real image of the IR PCB sensor that is designed and built for this project

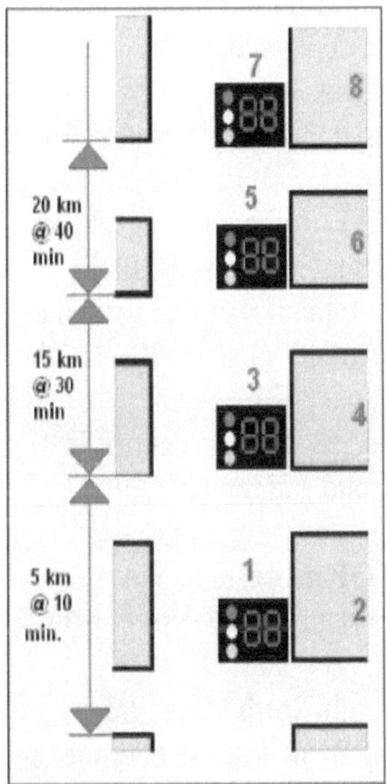

Figure 1.3

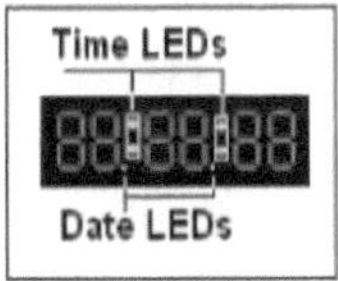

Figure 1.4

System Hardware

Introduction

1. A SIMATIC S7-300 PLC is the type of controller used in this project as the processor to control the traffic light. This type of PLC has been chosen because its characteristic fully matches what is necessary for the development of the traffic light system. See figure 1.6.

2. An **HMI** device made by **PanelMaster** is used to simulate the traffic lights on its screen and to send the START, STOP and manual **Emergency** signals or change the mode of operation from normal to Emergency when any pushbutton symbol is depressed.

3. The main control PCB is designed to be used as an interface between the PLC's I/O port terminals and all the signal lights and seven segment display boards. See figure 1.5 and 1.8.

4. Design of a 1×2 seven segment display PCB is to show timing and control of red, yellow and green stop lights that is connected to the main control PCB. In figure 1.5, notice that total of 9 of this secondary PCBs are used for each intersection and the tenth is used to display the speed (PCB # 10).

5. Design of a 1×6 digits PCB is also connected to the main control PCB to display time and date in hour/minute/second and month/day/yy format respectively (PCB # 9 in figure 2.1).

6. Three pushbuttons are used to **START**, **STOP** the system, and Emergency one is used to change the mode status of the system from normal to Emergency. See figure 1.5.

7. S1 and S2 IR sensors are installed on the South-North Main Street within 100 meters distance from each other. By measuring the time when a car passes both IR sensors, control program of PLC is used to calculate the speed at which the car is traveling and shows it on the 9th PCB.

8. From figure 1.15, notice that at 30 km/h it takes 10 minutes to get from first intersection to the second one, 30 minutes to get to the third one, and 40 minutes to get to the forth one. Any driver driving at 30 Km/h

will get a "green" signal light. This way, we can encourage drivers to drive within the speed limit.

9- Let's say it takes 5 seconds for a car to travel from S1 to S2. To calculate the speed of the car in Km/h, we know that if a car is traveling at 1km/h, it is traveling **1 kilometer every hour**. We can convert from m/s to km/h using this conversion equation: **1 m/s = 3.6** km/h.

To calculate the speed of the car for traveling 100 meters/5 second we can write:

$$100 \div 5 = 20 \ \mathbf{m/s} \times 3.6 \ \mathbf{km/h} = 72 \ \mathbf{km/h}$$

In the control software program, the speed limit is set to 30 km/h. Therefore, when a car travels 100 meters in 5 seconds, it is going over the speed limit by 72-30 = 42 km/hour driving faster than 30 km/h. The control program calculates car's actual speed which is 72 km/h and displays it on the speedometer display and it also turns the red LED on.

Figure 1.5 shows block diagram of the PLC system connected to the HMI, 3 pushbuttons, 2 IR sensors (S1 and S2), main PCB, and the main PCB's connection to all secondary seven segment displays.

Block diagram of the PLC system connected to HMI, main PCB etc

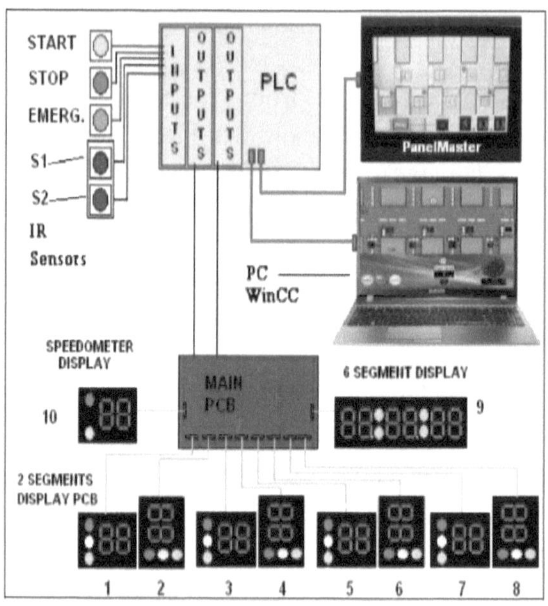

Figure 1.5

Figure 1.6 shows the S7-300 mounted on a rack. It includes the following components: 1- CPU: The central processing unit. 2- A Digital Input module 24 V DC transistor 3- Two Digital Output modules 24 V DC transistor 4- One Digital Output module (DO 16× AC/ 120/230 V).

Figure 1.7 displays the five input signals generated with the **START, STOP, Emergency, S1** and **S2** IR sensors connected to **Digital Input** terminals of S7-300 PLC. PLC's output terminals are connected to the **Main PCB** which is feeding all other seven segment display PCBs with data.

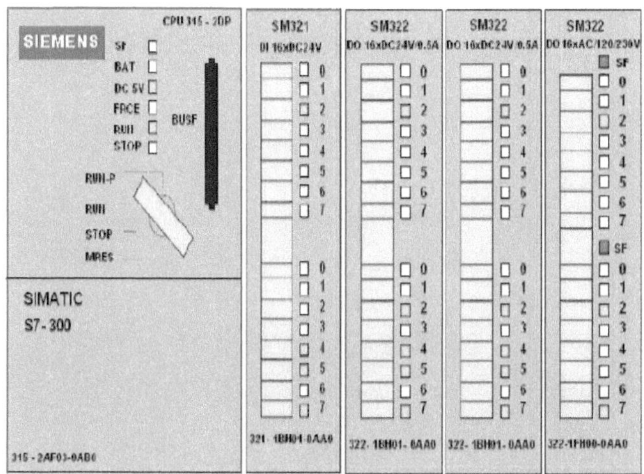

Figure 1.6 displays the PLC used to implement the project

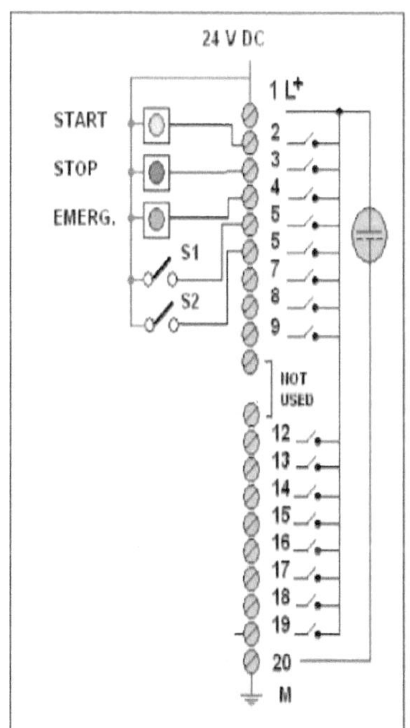

Figure 1.7

Figure 1.8 displays how PLC output terminals are connected to the Main PCB. Figure 1.9 displays the schematic diagram of electronic circuitry on the Main PCB.

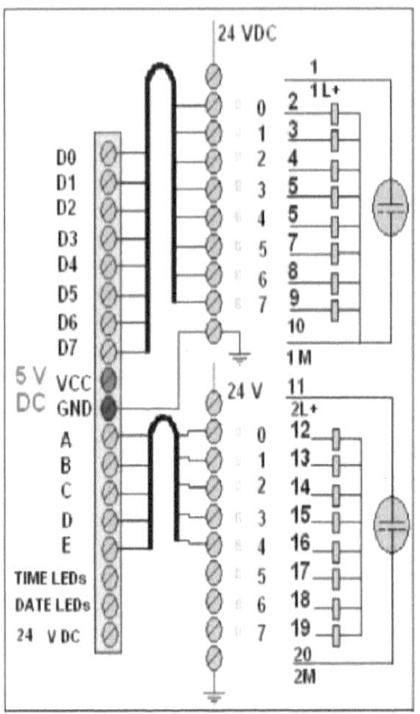

Figure 1.8

Notice that on main control board, we have 10 IDC each with 2 x 14 pins. Eight of these 2 x 14 pin headers are going to be connected to 8 seven segment display boards (countdown timers) plus each has 3 signal lights (red, yellow and green). The 9th display board is the speedometer PCB. Finally the 10th display PCB, shows the time and date. When 4 LEDs blinking, it is displaying time in hh:mm:ss format and when 2 of seven segment display points are on (solid), it is displaying date in (day, month, and year) format.

Figure 1.10 is related to one of the two digits display PCB. Notice by setting the address and signal setting jumpers, we can design one PCB and use it with any different address in more that one location. Therefore, in figure 1.9, by setting these jumpers, we are using the same PCB for nine different addresses. With the current setting, PCB with schematic shown in figure 1.10 is used as seven segment PCB display for the first intersection (S/N road).

Figure 1.9 displays the schematic diagram of electronic circuitry on the Main PCB.

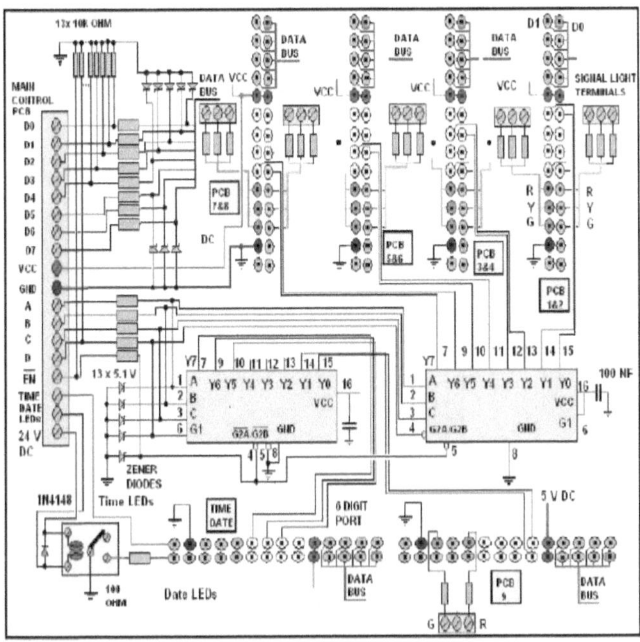

Figure 1.9 schematic diagram of main control PCB

Schematic diagram of 2 × 7-segment digits display PCB

Figure 1.10 displays wiring diagram of each 2 seven segment display board plus 3 LED indicators as traffic signal lights (red, yellow, and green).

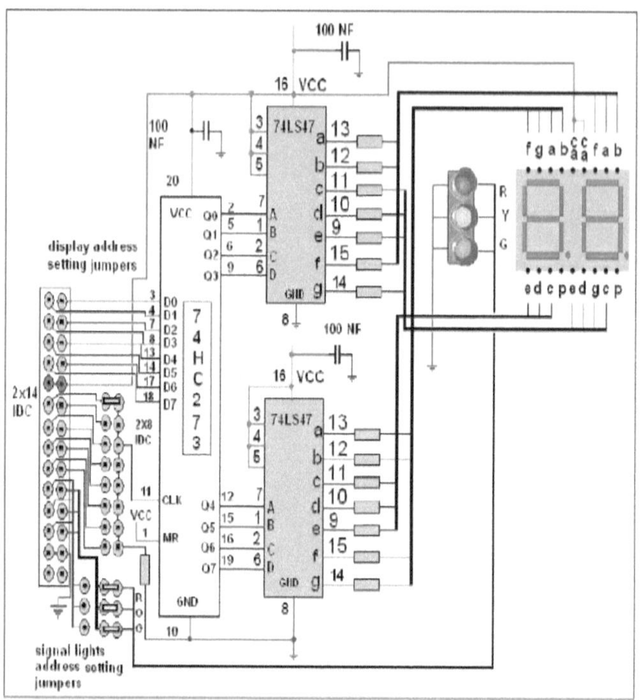

Figure 1.10 displays wiring diagram of each 2 seven segment display board plus 3 LED indicators as traffic signal lights (red, yellow, and green).

Figure 1.11 displays schematic diagram of TIME and DATE indicator PCB. It is actually 3 modules of figure 1.10 housed in one PCB. Figure 1.12 displays S7-300 3 × Output terminals which are used in this project to output data to the main control PCB.

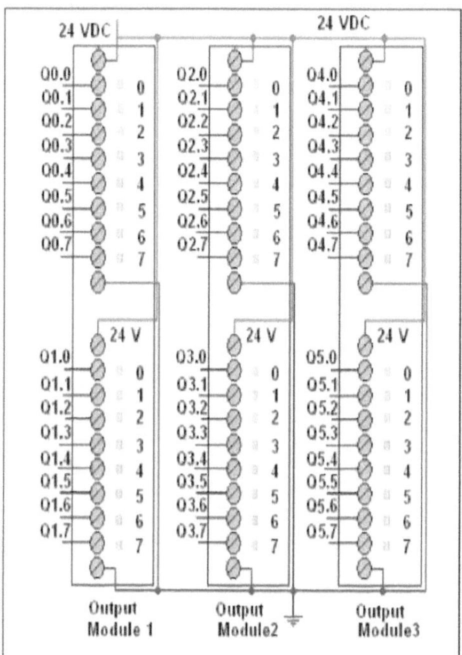

Figure 1.12

Figure 1.13 displays wiring schematic diagram of PLC output terminals to the main control PCB. From figure 1.13, notice that we need to use 8 × 3 = 24 output terminals to turn signal light on or off, 8 output terminals to write data on data bus, 5 output terminals to address data and latch, 2 terminals to control indicators on 6 seven segment display PCB (shows time and date of the current day) and 2 terminals to turn on the red or green LEDs regarding speedometer PCB to indicate if the car is over speeding (traveling speed > 30 km/h) or <= 30 km/h respectively.

Total of PLC Output terminals used = 24+ 8 + 5 + 2 + 2 = 41.

Comparing figure 1.13 with 1.14, you should be able to easily relate the signal light locations to the output terminal port addresses.

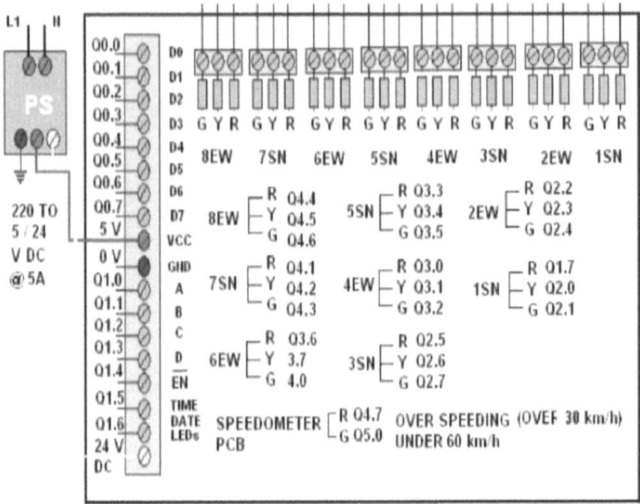

Figure 1.13

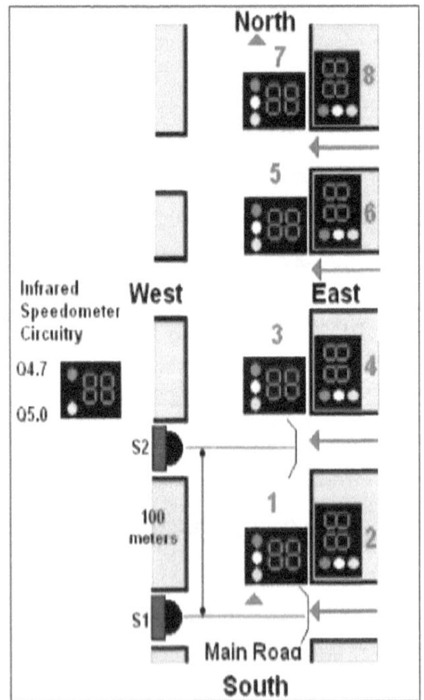

Figure 1.14

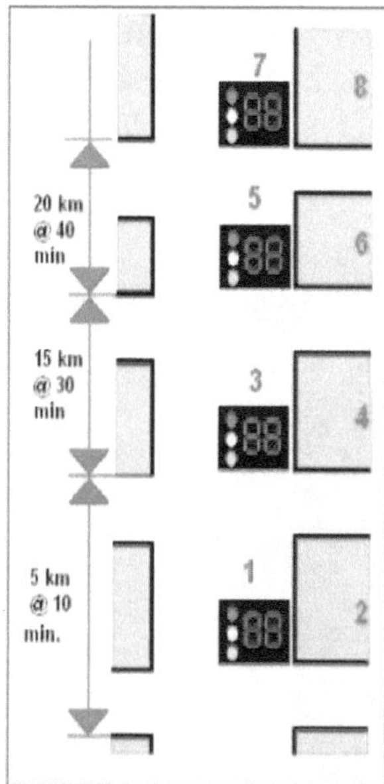

Figure 1.15

From figure 1.5, notice that an HMI device is used to monitor all traffic lights and countdown timers and other input or output activities. For example, when the IR beam breaks, by a car passing by the S1 and S2, the PLC calculates and displays the passing car's speed. In this case, HMI also displays the speed. See figure 1.16.

Figure 1.16 displays HMI hardware used in this project

Figure 1.17 displays the third software program developed for this project that simulates the main intersection streets, all traffic lights, countdown timers, pushbuttons, current time and date displays and transfers all these data to a related PC's monitor.

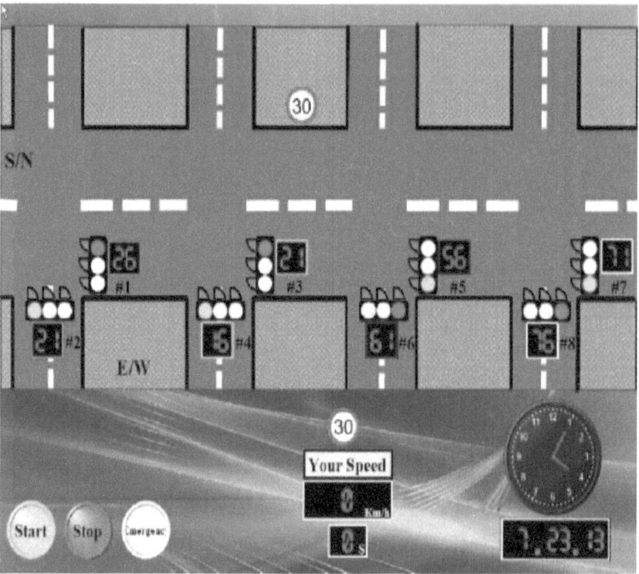

Figure 1.17

Developing main control software for S7-300 PLC

In this chapter, we are going to develop the main control ladder logic program for S7-300 PLC which when executed, it will get all display PCBs activated and all input devices such as pushbuttons, IR sensors, etc being connected to the related input terminals, cause PLC to receive and execute the input signals.

Full explanation of the project software is given in chapter 2. You are assumed to be able to develop a circuit diagram of this project with even extra features on your own. Just in case, if you wish to have the code developed for this project, you only need to email me proof of your purchase form a website, eBay, Amazon etc to my email address srfattah@yahoo.com. Upon receipt of that proof, the software will be sent to your email address. The following table shows the folder and file names that I used to develop my circuit program or the application programs for this project:

Program description	Folder name	File name
Main PLC program	Main_CP	Ch_ra.s7p
HMI settings	HMIN Program	v1.pm2
Wincc settings	Traffic_light_wincc	Stop_light.mcp

To see the project in action, please click on the following link:
http://youtu.be/0ADfFPOzIUE

Phase 1 – developing the PLC main ladder logic program (Ch_ra.s7p)

Figures 2.1 and 2.2 display the assignment of different field devices or I/O symbols to addresses. According to figure 2.1, notice that **Q0.0 to Q0.7** outputs are assigned to D0 to D7 respectively. By these outputs, we are going to send any two digit numbers to any two seven segments to display different information, such numbers being decreased by one unit every second on our timer displays at each intersection.

Symbol Editor - [S7 Program(2) (Symbols) -- ch_ra\SIMATIC 300 Station\CPU 315-2 DP]

Symbol Table Edit Insert View Options Window Help

Al Symbols

	Status	Symbol	Address		Data type	Comment
1		A	Q	1.0	BOOL	
2		B	Q	1.1	BOOL	
3		C	Q	1.2	BOOL	
4		D	Q	1.3	BOOL	
5		D0	Q	0.0	BOOL	
6		D1	Q	0.1	BOOL	
7		D2	Q	0.2	BOOL	
8		D3	Q	0.3	BOOL	
9		D4	Q	0.4	BOOL	
10		D5	Q	0.5	BOOL	
11		D6	Q	0.6	BOOL	
12		D7	Q	0.7	BOOL	
13		emergency	I	0.2	BOOL	
14		on	Q	1.4	BOOL	
15		g_113	Q	2.1	BOOL	
16		g_123	Q	2.4	BOOL	
17		g_213	Q	2.7	BOOL	
18		g_223	Q	3.2	BOOL	
19		g_313	Q	3.5	BOOL	
20		g_323	Q	4.0	BOOL	
21		g_413	Q	4.3	BOOL	
22		g_423	Q	4.6	BOOL	
23		L_speed1	Q	4.7	BOOL	
24		L_speed2	Q	5.0	BOOL	
25		LD	Q	1.6	BOOL	

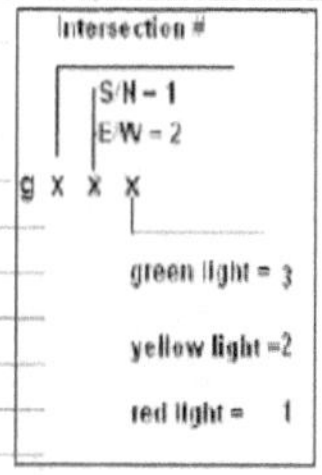

Figure 2.1

26		LT	Q	1.5	BOOL	
27		r_111	Q	1.7	BOOL	
28		r_121	Q	2.2	BOOL	R1
29		r_211	Q	2.5	BOOL	
30		r_221	Q	3.0	BOOL	
31		r_311	Q	3.3	BOOL	
32		r_321	Q	3.6	BOOL	
33		r_411	Q	4.1	BOOL	
34		r_421	Q	4.4	BOOL	
35		READ_CLK	SFC	1	SFC 1	Read System Clock
36		speed1	I	0.3	BOOL	
37		speed2	I	0.4	BOOL	
38		start	I	0.0	BOOL	
39		stop	I	0.1	BOOL	
40		TOD_INT0	OB	10	OB 10	Time of Day Interrupt 0
41		y_112	Q	2.0	BOOL	
42		y_122	Q	2.3	BOOL	
43		y_212	Q	2.6	BOOL	
44		y_222	Q	3.1	BOOL	
45		y_312	Q	3.4	BOOL	
46		y_322	Q	3.7	BOOL	
47		y_412	Q	4.2	BOOL	
48		y_422	Q	4.5	BOOL	
49						

Figure 2.2

Phase 2

At phase 2, we develop network #1 related to OB100 to reset all bits that might have been previously set. See figure 2.3.

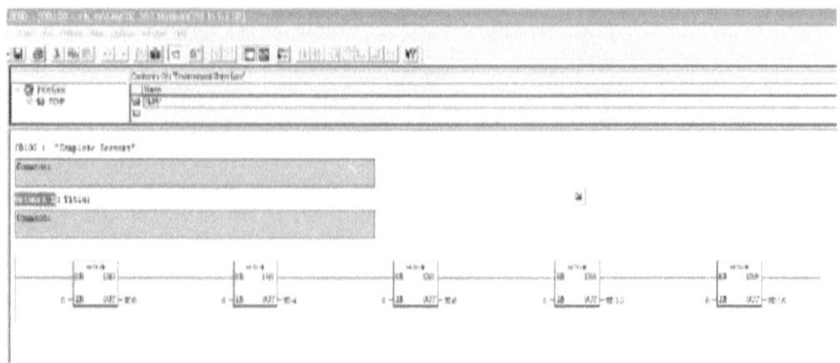

Figure 2.3

Phase 3

Now we go head and develop codes related to the Emergency, START, and STOP pushbuttons. First for the Emergency pushbutton.

Network 173 : Title:

Comment:

```
    I0.2
"emergency
 "                                          M130.1
  ┤ ├                                      ─( S )─┤
```

Figure 2.4

Emergency mode, is activated either by depressing **Emergency** pushbutton (figure 2.4), or at 24.00 PM, every day. At 7.00 AM, system automatically, maintains its normal mode. We can use Time of Day Interrupt to get the system to go into its **Emergency mode** (at 24:00 and back to its normal operation at 7:00 AM). To fix the needed setting, double click on CPU to get **Properties –CPU 315-2 DP**. (Yours might be a different CPU type). Dialog box will appear and do the setting as shown in figure 2.5.

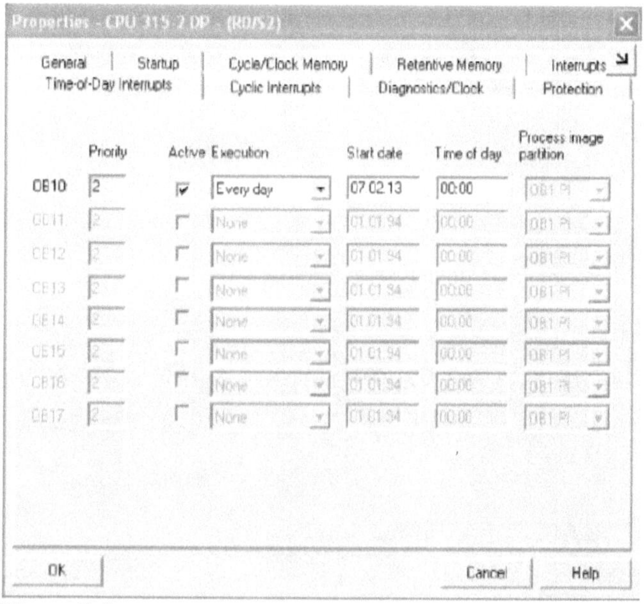

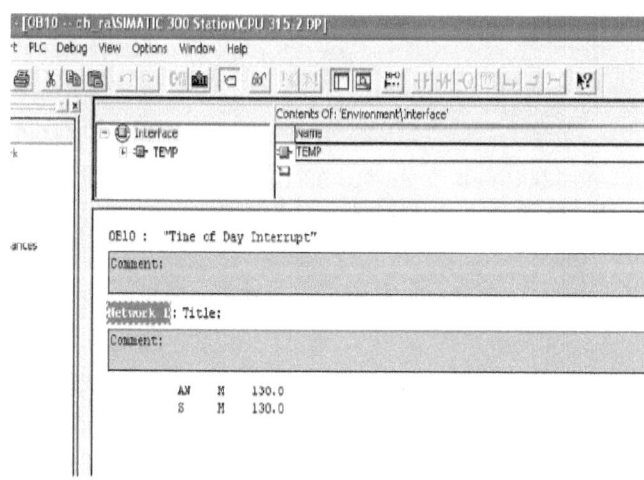

Figure 2.5

With respect to the setting done, PLC executes network from OB10 every day at 24:00 PM. Hence, we include our program in **OB10**. See figure 2.6.

Figure 2.6

And to reset the related flag for the Emergency case, we can use the following network to get the flag reset at 7:00 AM.

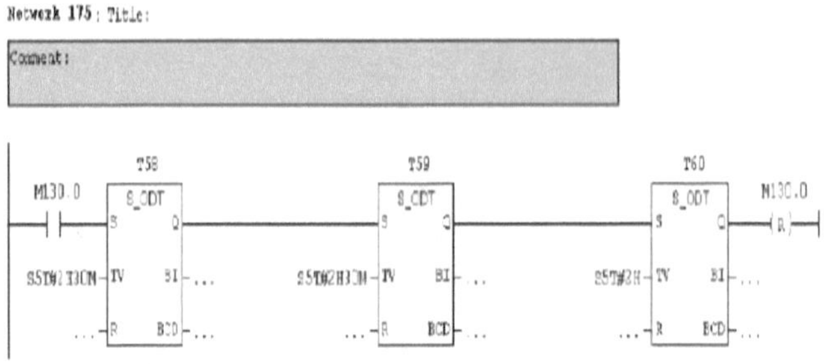

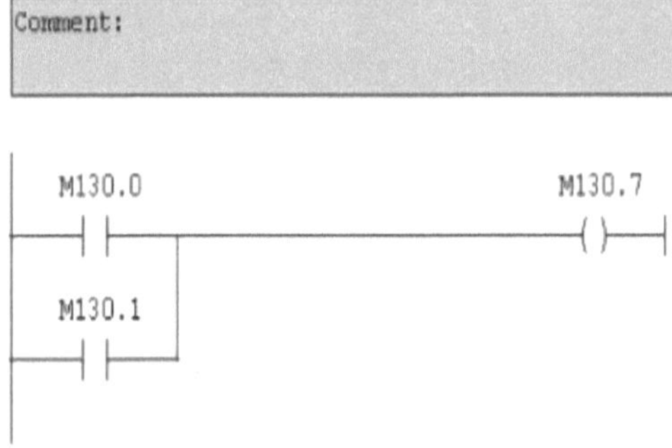

Figure 2.7

In either manual or Automatic Emergency case, M130.7 is set to 1. See <u>figure 2.8</u>

Figure 2.8

Also, when the START pushbutton is depressed, its related flag must be SET, and the one related to the Emergency one, must be RESET.

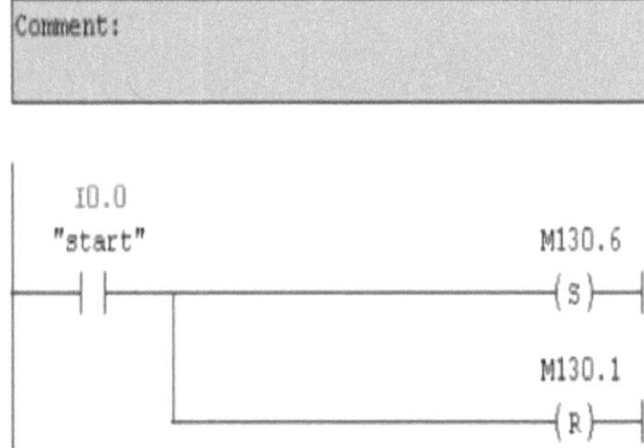

Figure 2.9

Also, for STOP pushbutton, the program should be written such to RESET all flags and outputs.

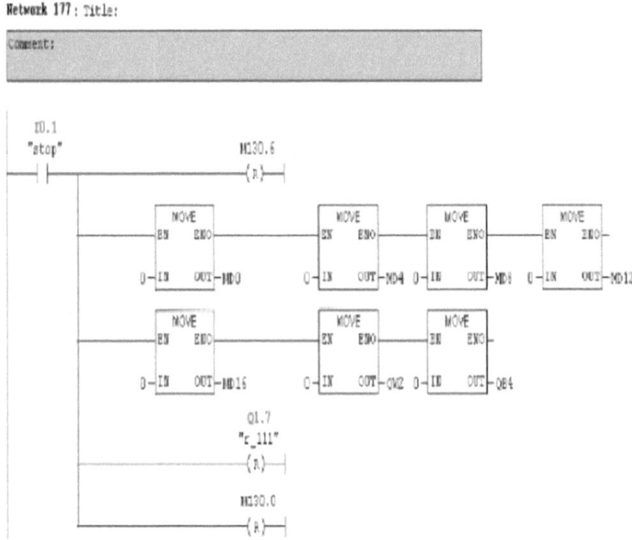

Figure 2.10

Phase 4

On this phase, all values of times are written in flags to be used when ever needed.

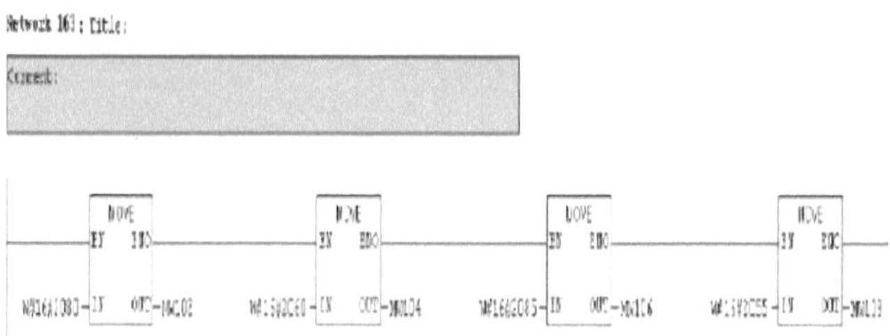

Figure 2.11

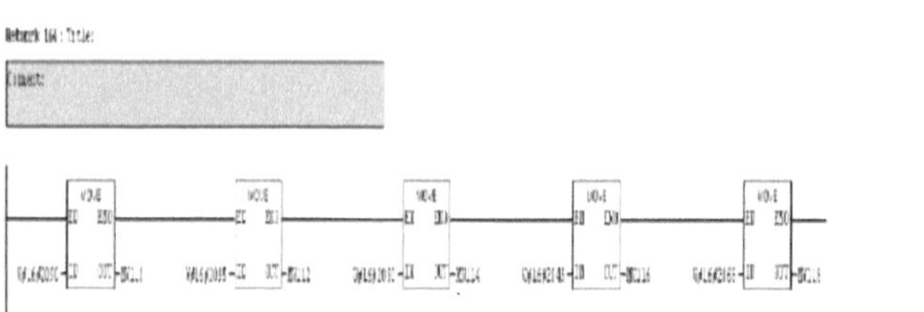

Figure 2.12

Phase 5

On this phase, the networks related to the START pushbutton is developed. When the START pushbutton is depressed, all flags and outputs related to the first intersection (it might be in Emergency mode too) are **RESET** and **Flags** related to the first intersection are turned on.

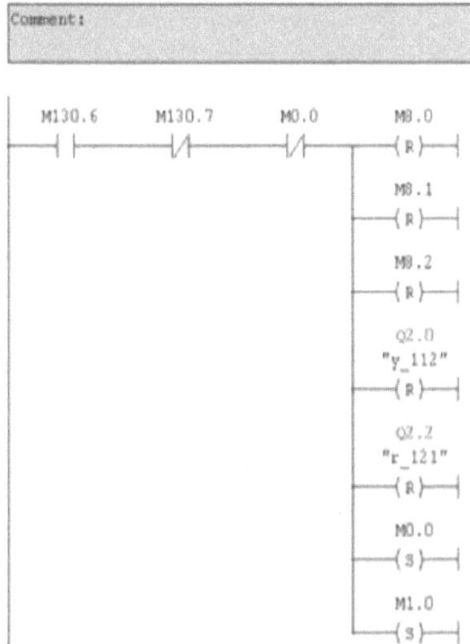

Figure 2.13

And the execution the program, turns on the first green LED (green stop light).

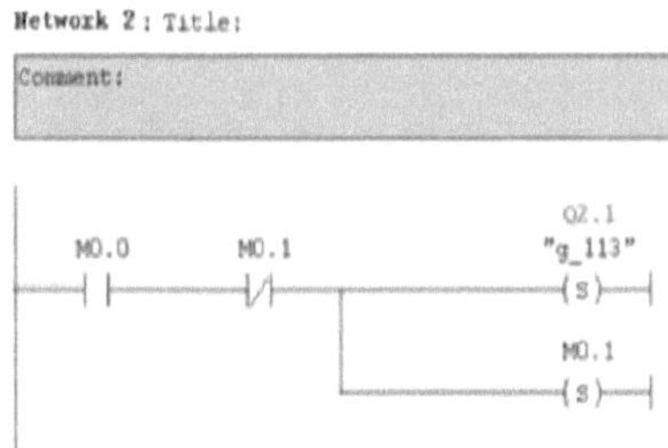

Figure 2.14

The first green stop light must be turned on and stay on for 80 seconds. We can use a timer to fix the time and save its value as a BCD number in a defined memory location.

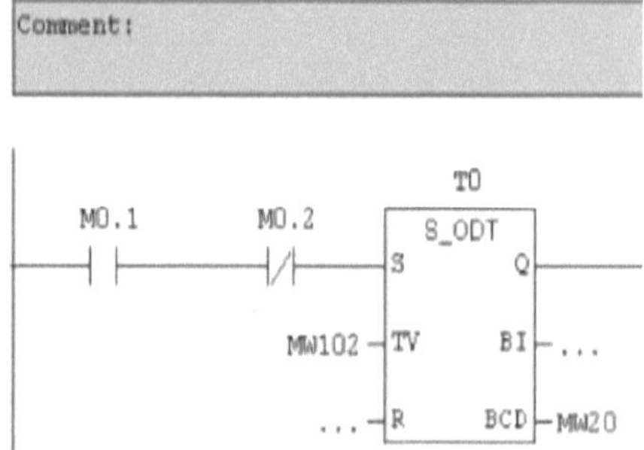

Figure 2.15

After 80 seconds, the green LED turns off, and the yellow one must be on for 5 seconds. And after 5 seconds, it must also be turned off, and the red LED (traffic light) comes on. The following codes can take care of these timing.

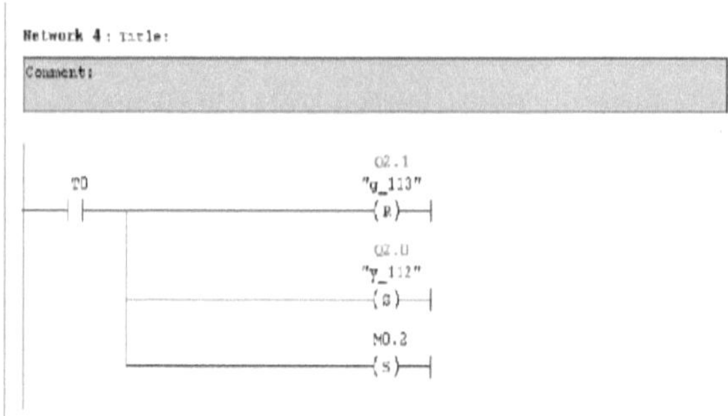

Figure 2.16

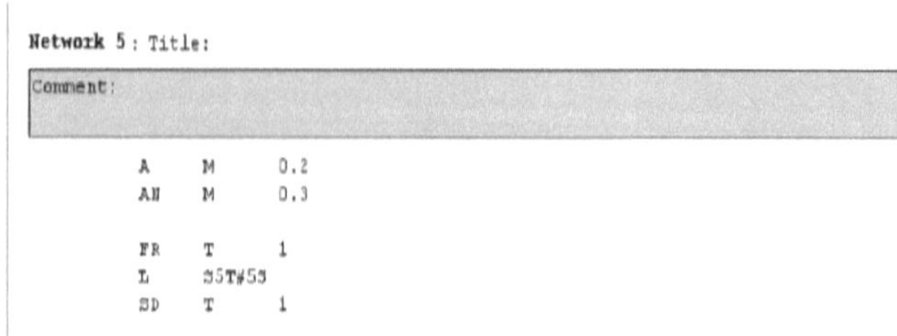

Figure 2.17

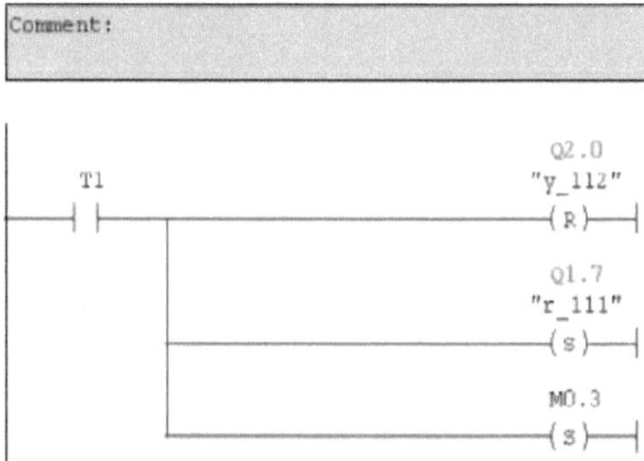

Figure 2.18

The red light stays on for 60 seconds. After that, all flags related to the light must be reset to make the condition satisfied for network 2 to be executed after which the execution of the program enters the continuation loop and as long as either the **Emergency** or **STOP** pushbutton is not depressed, the execution of the program repeats itself.

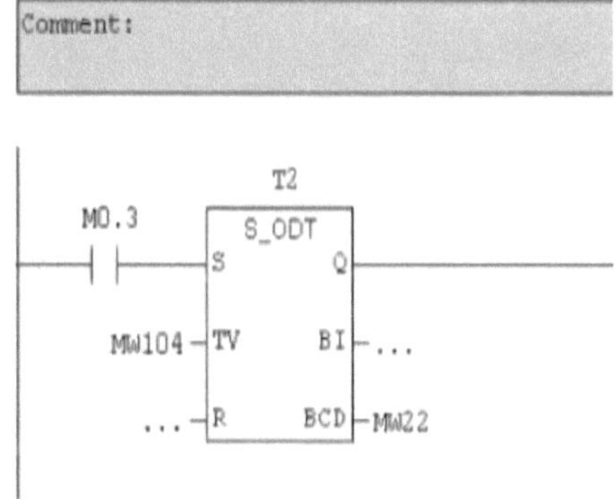

Figure 2.19

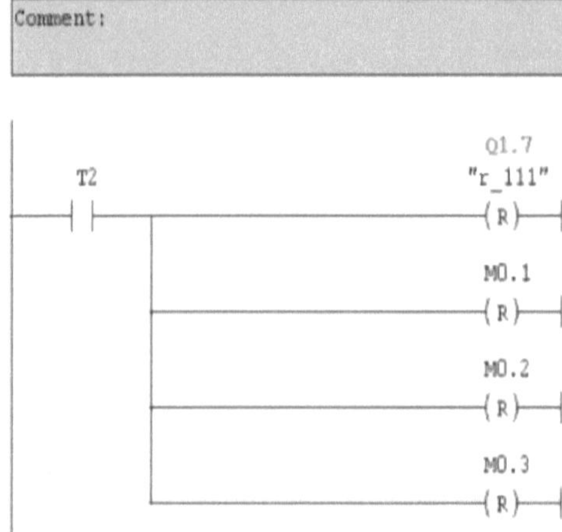

Figure 2.20

On the other hand, the lights on the **E/W Street** (PCB # 2) turn on with the opposite colors. Based on figure 1.1, R2 = 85, Y2 = 5 and G2 = 55 seconds. The coding for this part is similar to the previous one.

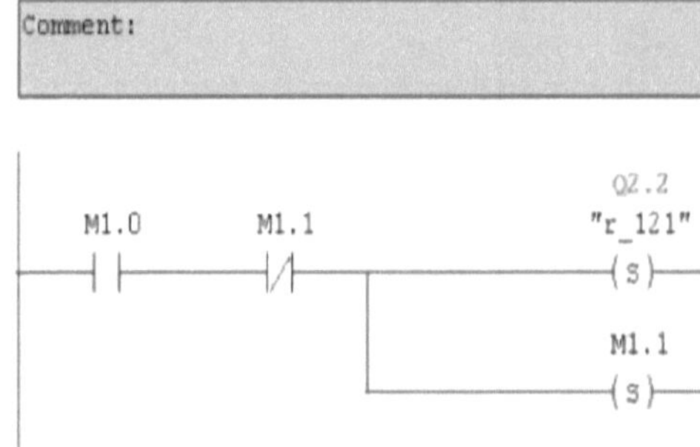

Figure 2.21

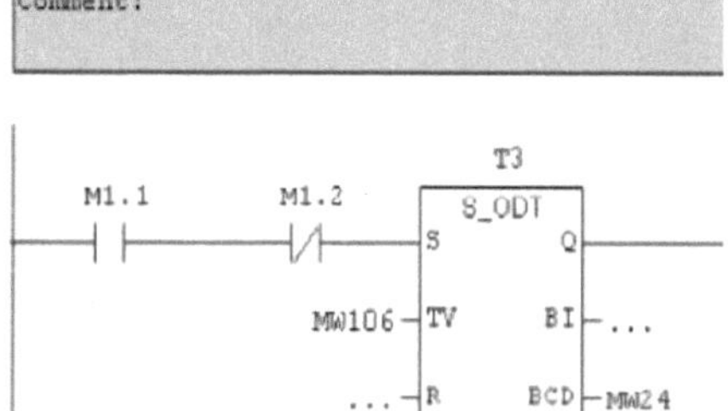

Figure 2.22

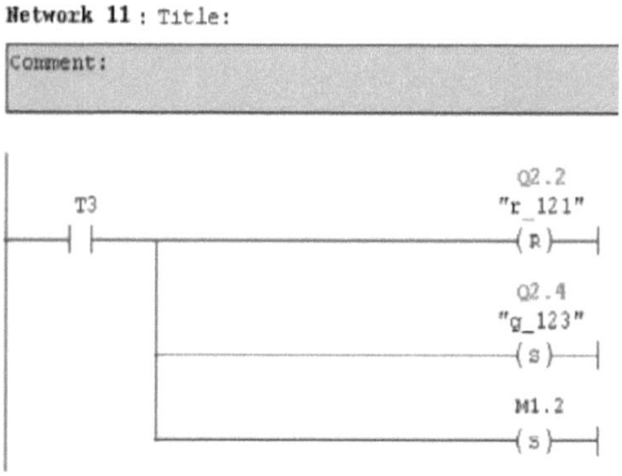

Figure 2.23

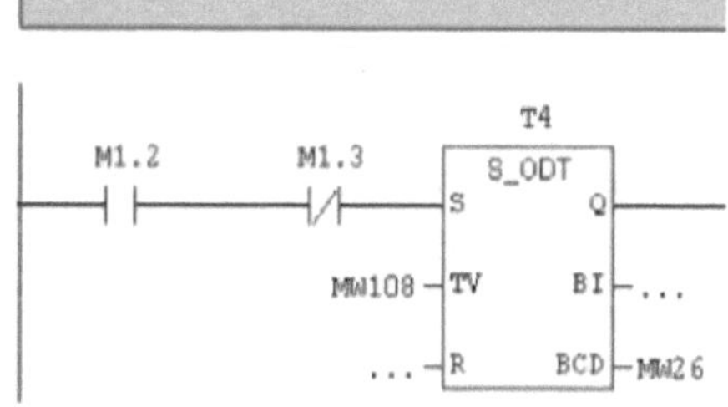

Figure 2.24

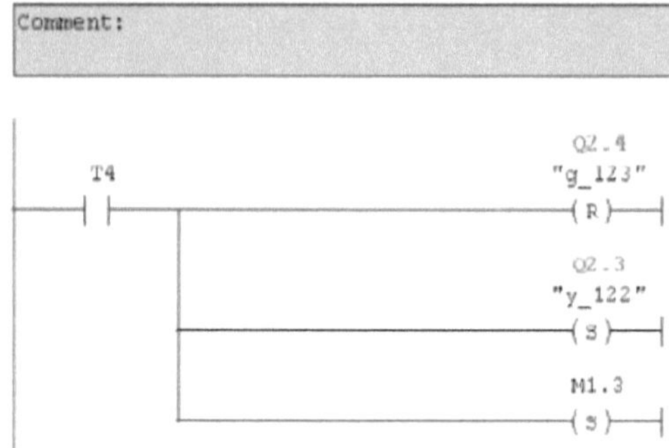

Figure 2.25

Network 14 : Title:

Comment:

```
A    M      1.3
FR   T      5
L    S5T#5S
SD   T      5
```

Figure 2.26

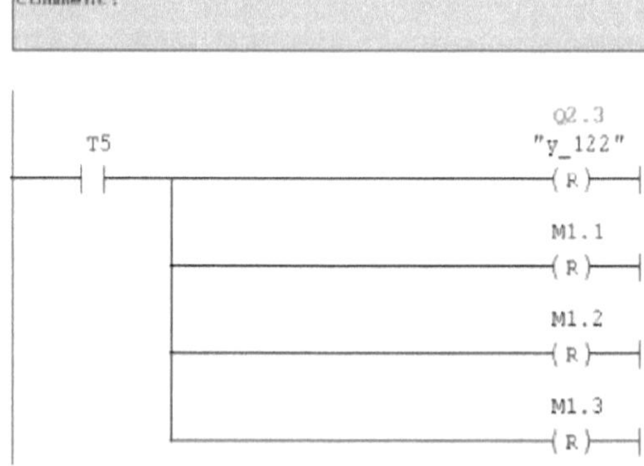

Figure 2.27

Phase 6 - Coding for the second intersection

On this phase, we need to take care of the coding related to the second intersection traffic lights. The distance between these two intersections is 5

Km. the speed limit at S/N road is 30 Km/h. To encourage drivers to drive within the speed limit, we would say if drivers stay within speed limit of 30 Km/h, to get to the second intersection, it will take 600 seconds. Based on the speed formula 5 Km = 30 Km/h × Time (h) Time = 5 ÷ 30 **Time = 10 minutes**. 10 minutes × 60 (seconds) = 600 seconds. Total amount of time for one cycle is = 115 seconds (in the second intersection). So coding should be prepared in such a way that second green traffic light turns on after 25 seconds after passing the first intersection traffic light (the green one). 600 = (5 × 115 + 25) = (575+25). This means that at the second intersection, the green traffic light turns on with a delay of 25 seconds right after the first one turns on.


```
      A     M     130.6
      AN    M     130.7
      AN    M       2.0

      FR    T       6
      L        S5T#25S
      SD    T       6
```

Figure 2.28

Second traffic light turns on after 25 seconds after the first one turns on.

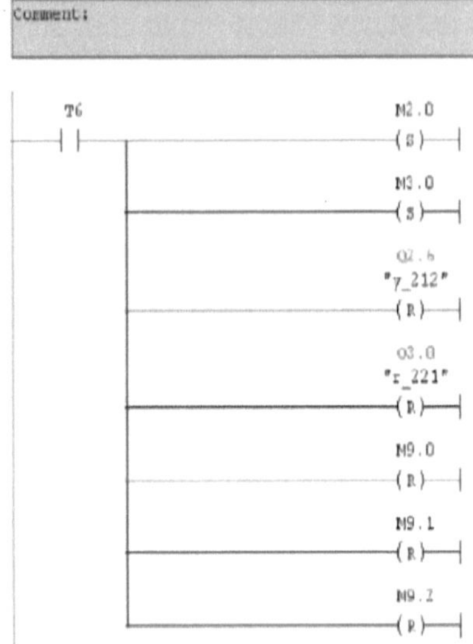

Figure 2.29

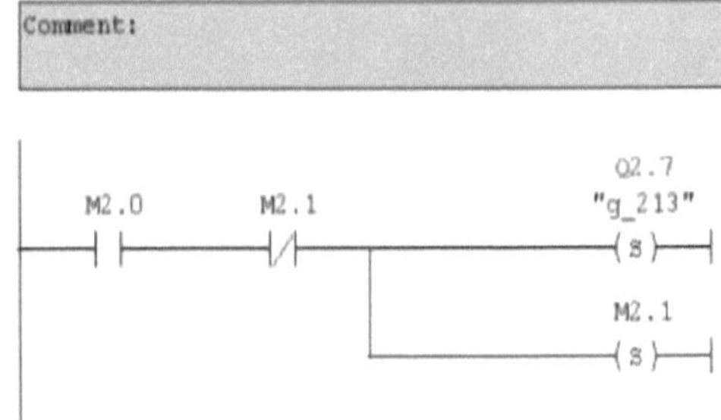

Figure 2.30

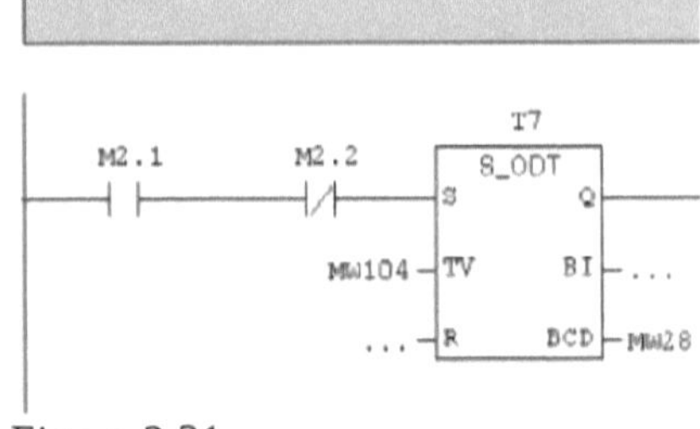

Figure 2.31

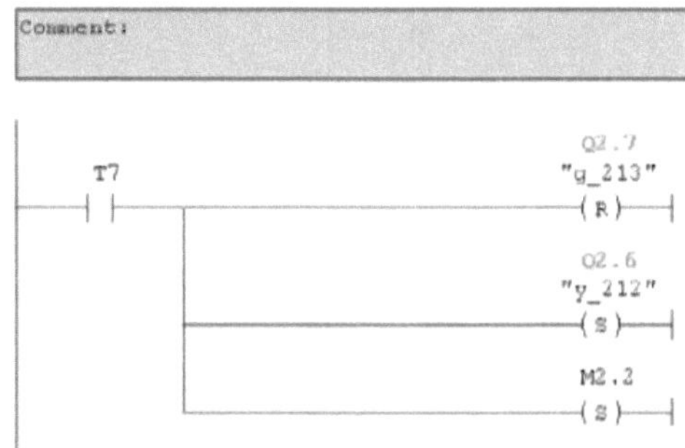

Figure 2.32

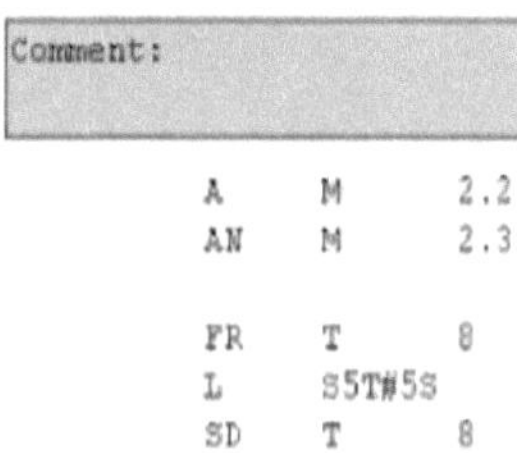

Figure 2.33

Figure 2.34

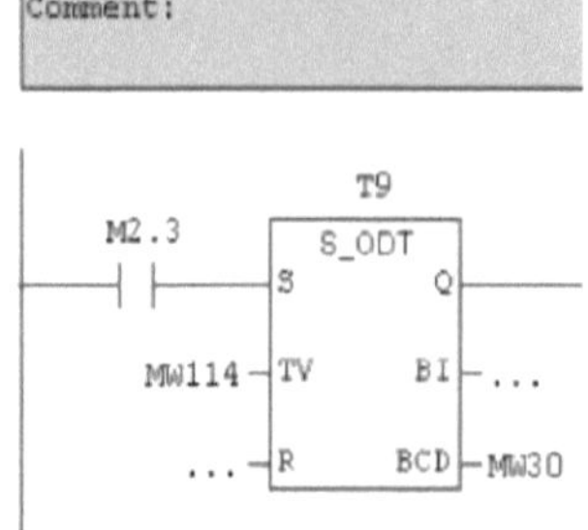

Figure 2.35

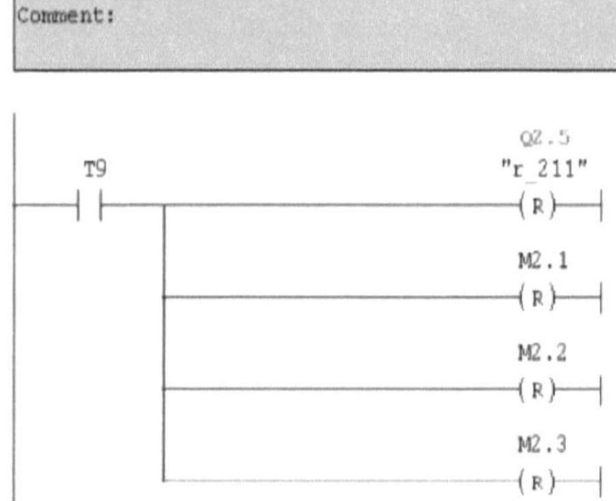

Figure 2.36

Coding to take care of the traffic lights R4 = 65 seconds, Y4 = 5 and G4 = 45 seconds is as following:

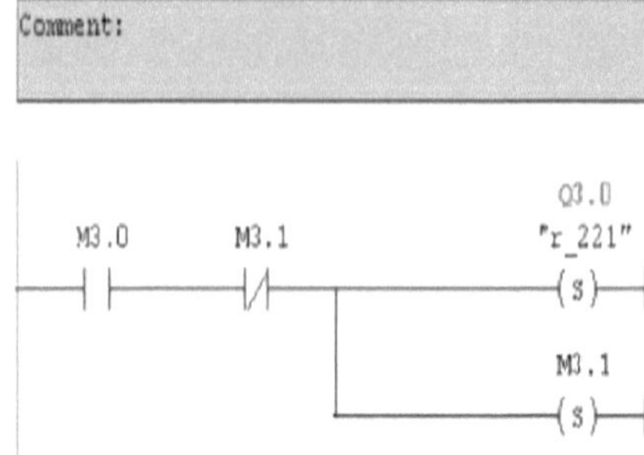

Figure 2.37

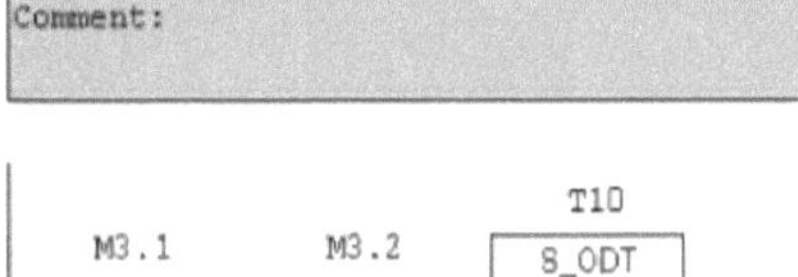

Figure 2.38

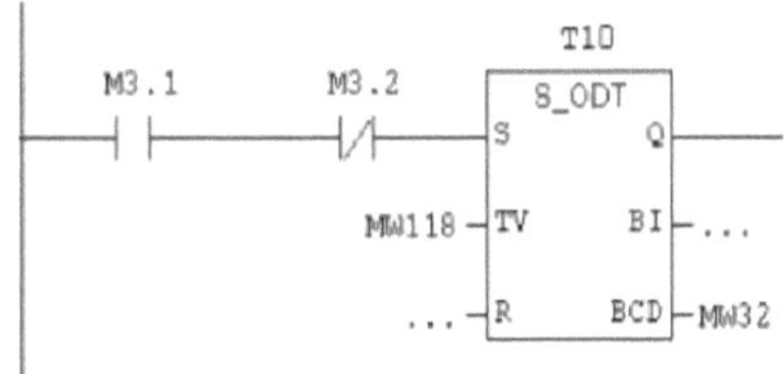

Figure 2.39

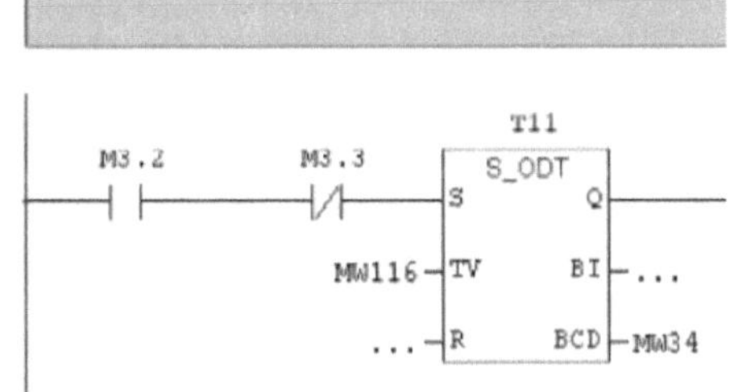

Figure 2.40

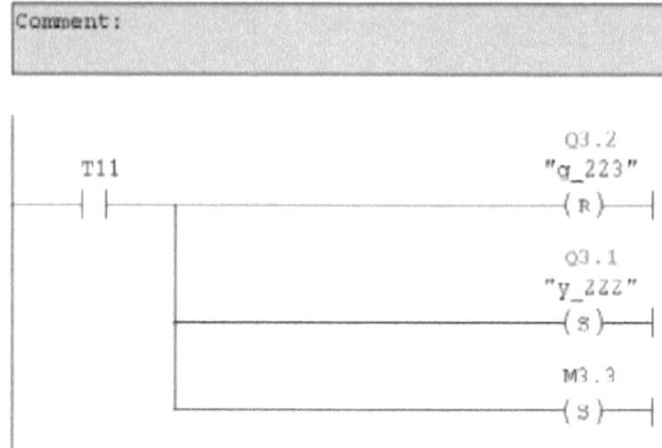

Figure 2.41

Network 30 : Title:

Comment:

```
A    M     3.3
FR   T    12
L    S5T#5S
SD   T    12
```

Figure 2.42

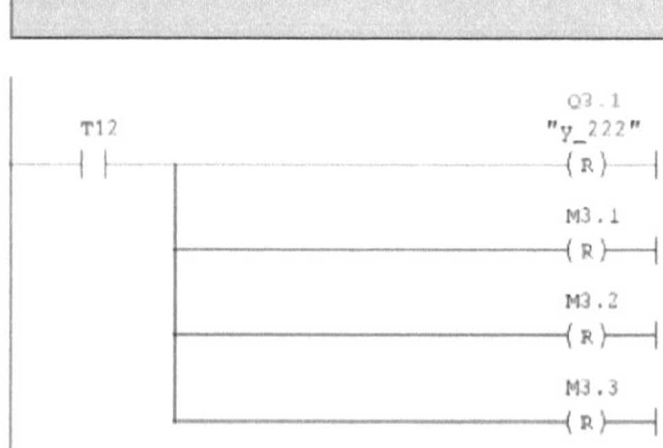

Figure 2.43

Phase 7 – Coding for the third intersection

On this phase, we need to generate coding for the third intersection's lights. From figure 1.1, these lights are on for total cycle time of **(g5 = 90, y5 = 5 and r5 = 50 seconds) 145** seconds. And the distance between these two intersections (second and third one) is 15 Km. So, it will take 30 minutes or **30 × 60 = 1800** = (12 × 145+60) = (1740 + 60) seconds. So, when the program in being **RUN** for the first time, after the second intersection's traffic light change to green, G5 turns on after 60 seconds. The following is the coding to execute what was said about the timing at the third intersection.


```
        A     M      2.0

        AN    M      4.0
        AN    M      5.0

        FR    T      13
        L     S5T#1M
        SD    T      13
```

Figure 2.44

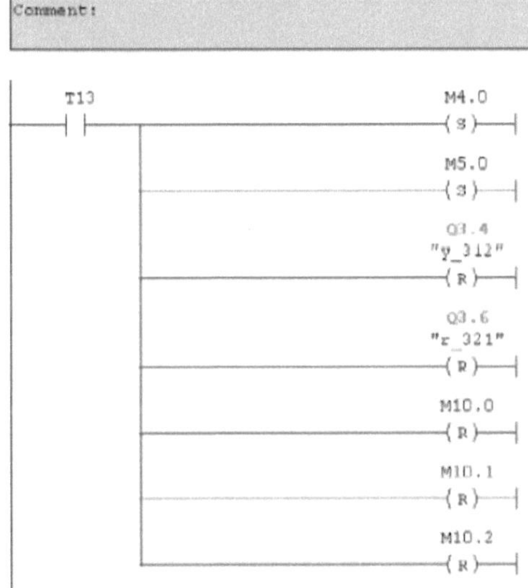

Figure 2.45

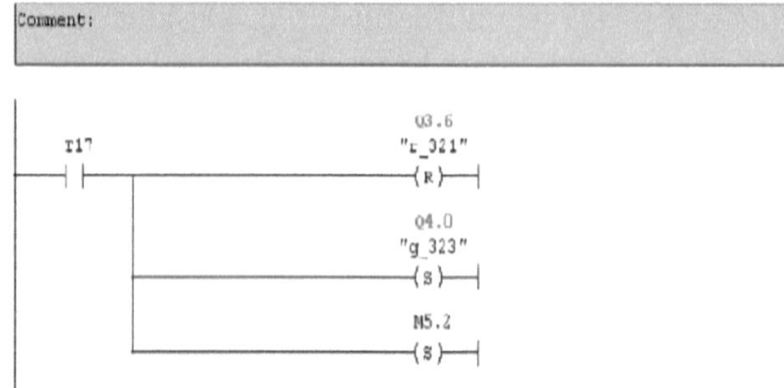

Figure 2.46

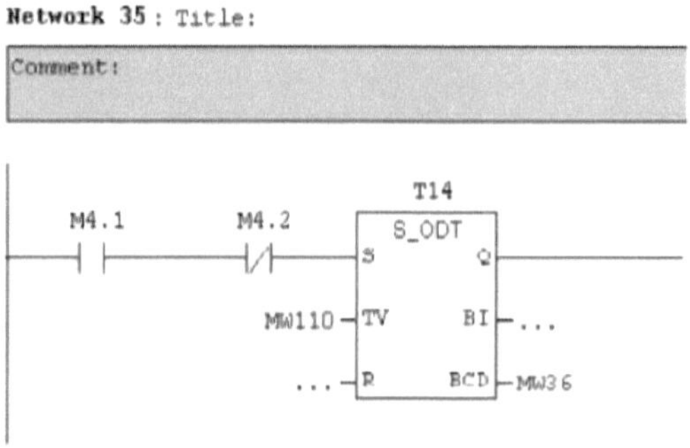

Figure 2.47

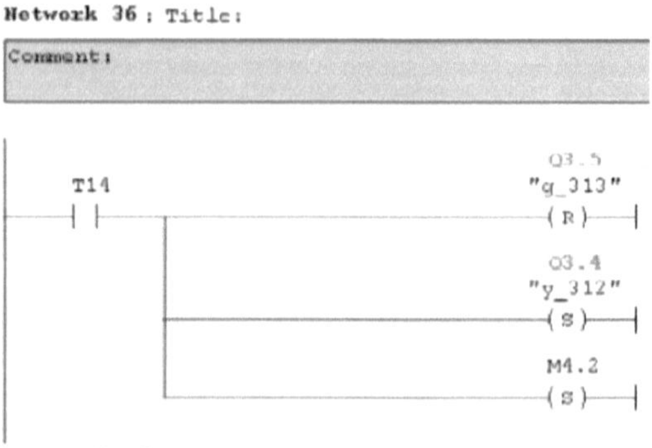

Figure 2.48

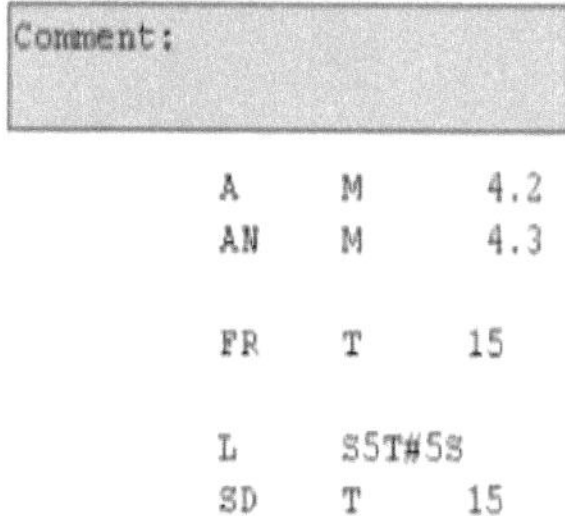

Figure 2.49

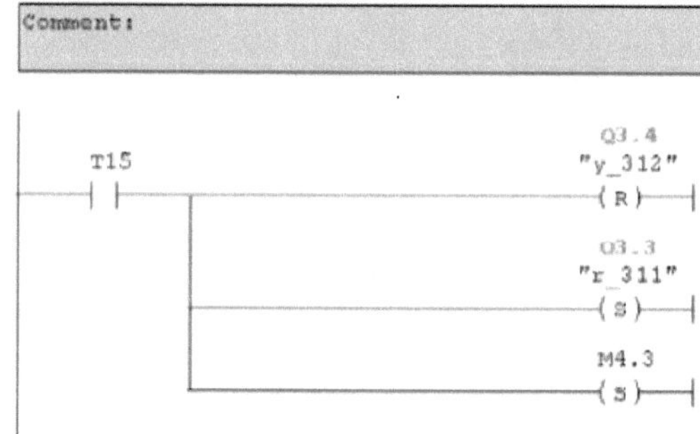

Figure 2.50

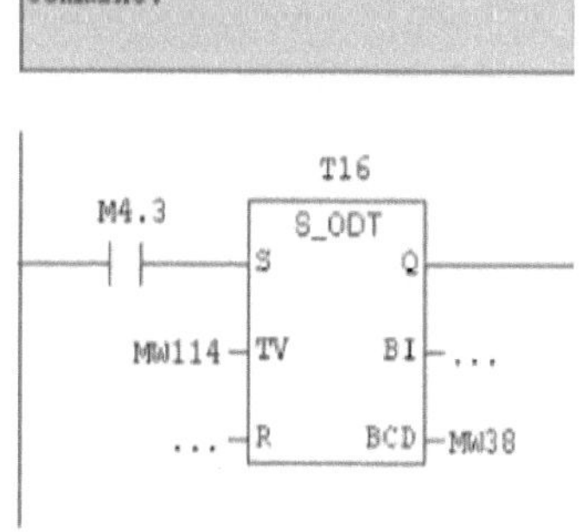

Figure 2.51

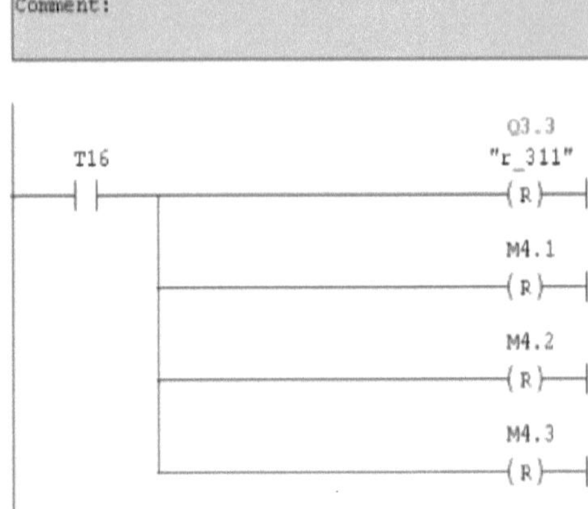

Figure 2.52

The coding for the **E/W** road related to R6 = 95, Y6 = 5 and G6 = 45 seconds is as following:

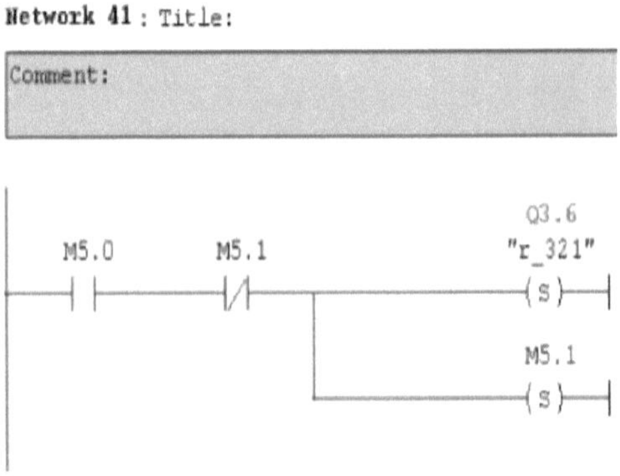

Figure 2.53

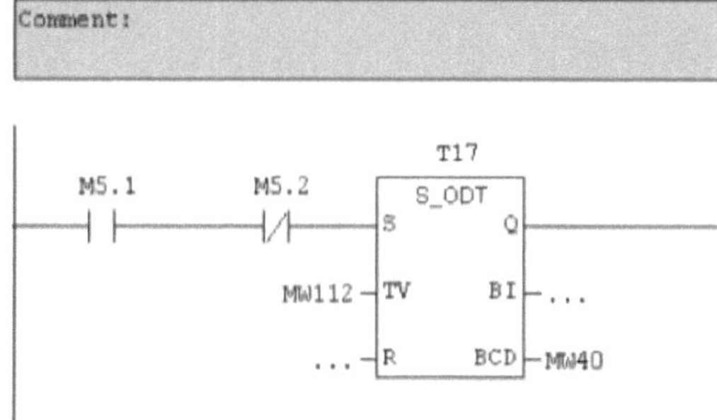

Figure 2.54

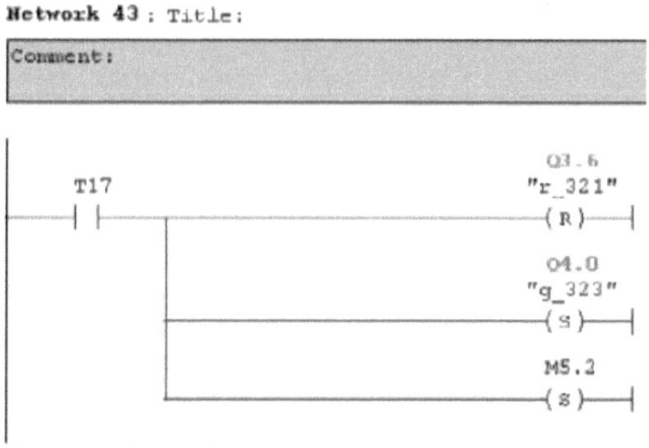

Figure 2.55

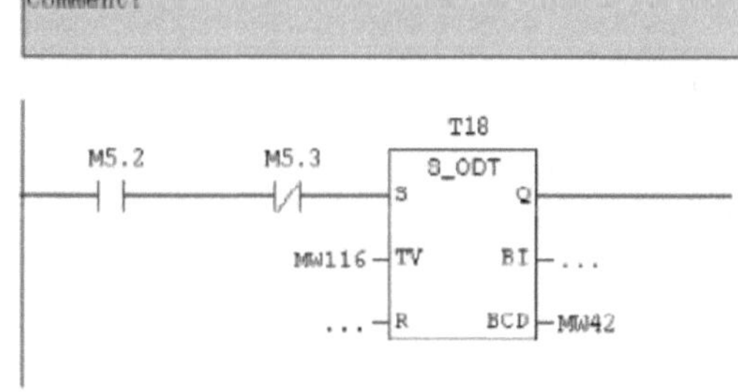

Figure 2.56

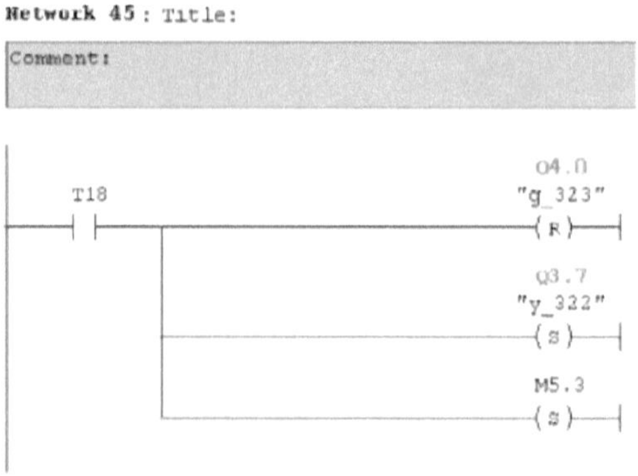

Figure 2.57

Network 46 : Title:

Comment:

```
A      M      5.3
FR     T      19
L      S5T#5S
SD     T      19
```

Figure 2.58

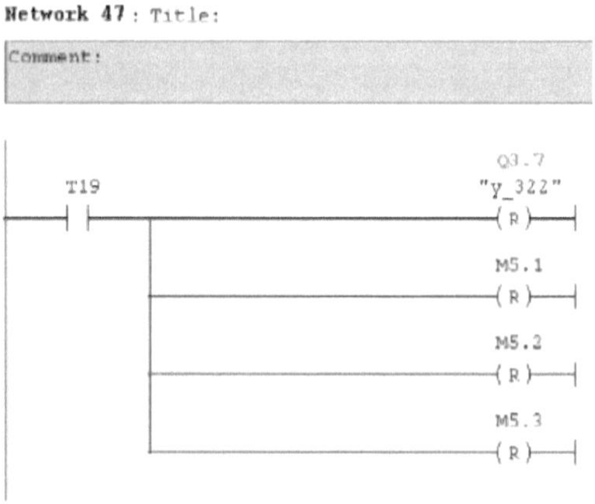

Figure 2.59

Phase 8 – Coding for the forth intersection

On this phase, we need to generate coding for the forth intersection's traffic lights. From figure 1.1, these lights are on for the total cycle time of

(g7 = 85, y7 = 5 and r7 = 50 seconds) 140 seconds. And the distance between these two intersections (second and third one) is **20 Km** and based on the calculation done earlier, it will take 40 minutes or **40 × 60 = 2400** = (17 × **140 + 20) = (2380 + 20)** seconds (with speed limit of 30 Km/h) for a driver to reach from the second intersection to the third one. So, when the program in being **RUN** for the first time, after the third intersection's traffic light changes to green, forth intersection's traffic light turn on and starts after 20 seconds and starts its **140** second cycle time.

```
Network 48 : Title:

Comment:

      A    M       4.0
      AN   M       6.0
      AN   M       7.0

      FR   T     20
      L    S5T#20S
      SD   T     20
```

Figure 2.60

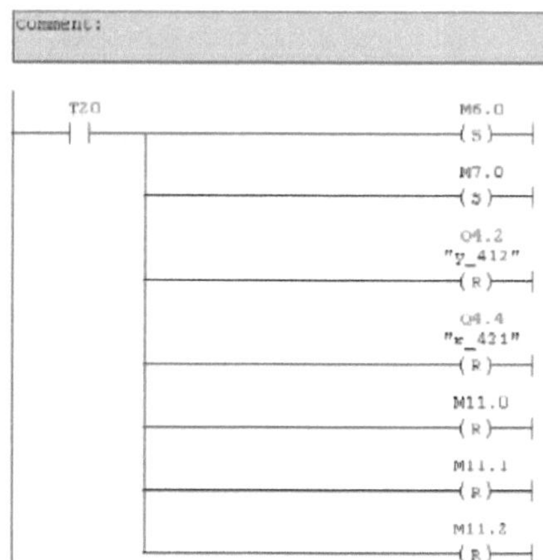

Figure 2.61

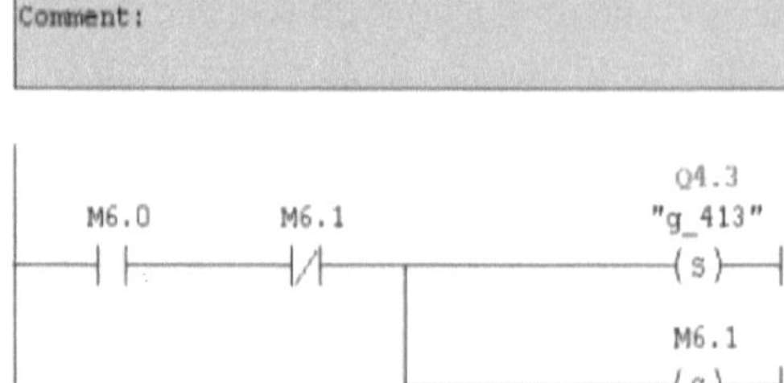

Figure 2.62

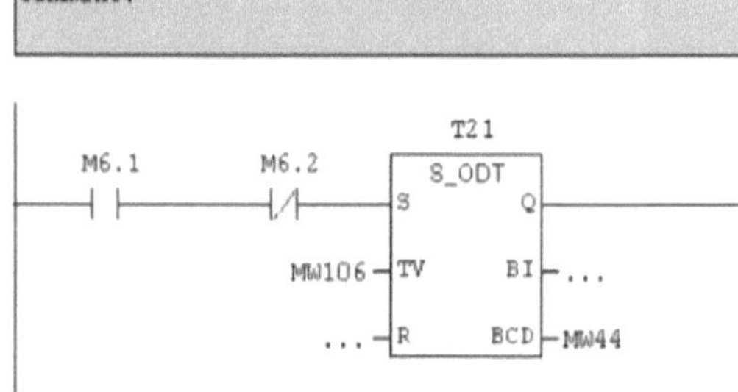

Figure 2.63

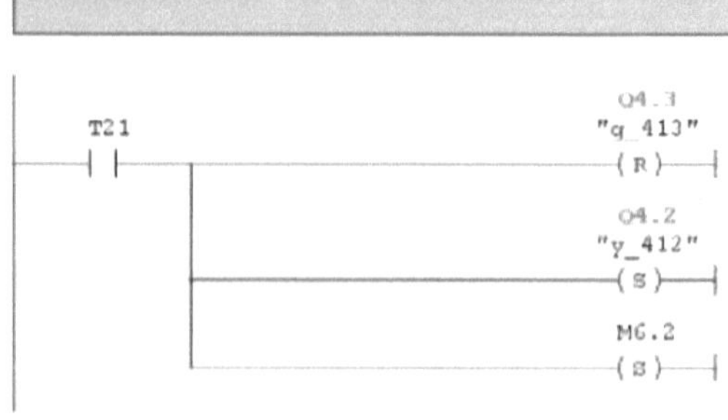

Figure 2.64

Network 53 : Title:

Comment:

```
A     M     6.2
AN    M     6.3

FR    T     22
L     S5T#5S
SD    T     22
```

Figure 2.65

Network 54 : Title:

Comment:

Figure 2.66

Network 55 : Title:

Comment:

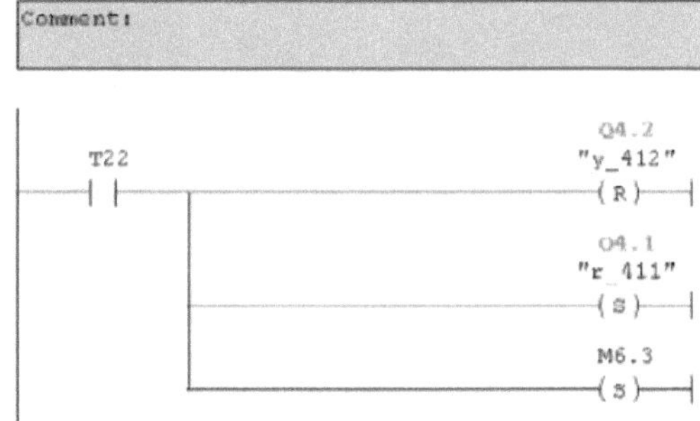

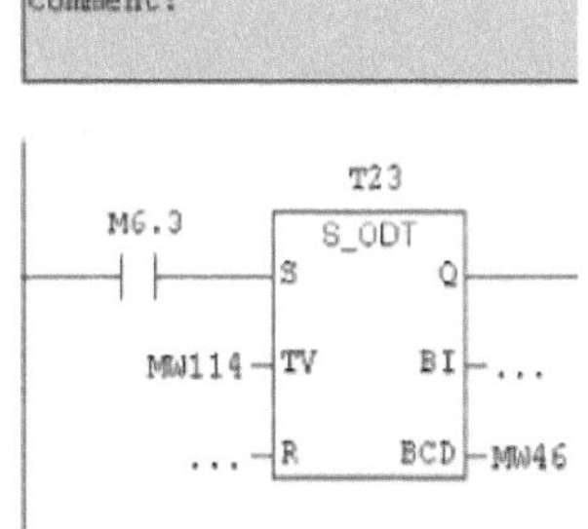

Figure 2.67

Figure 2.68

Now the generation of coding for the last E/W road with R8 = 90, Y8 = 5 and G8 = 45 seconds. The total cycle time is also 140 seconds.

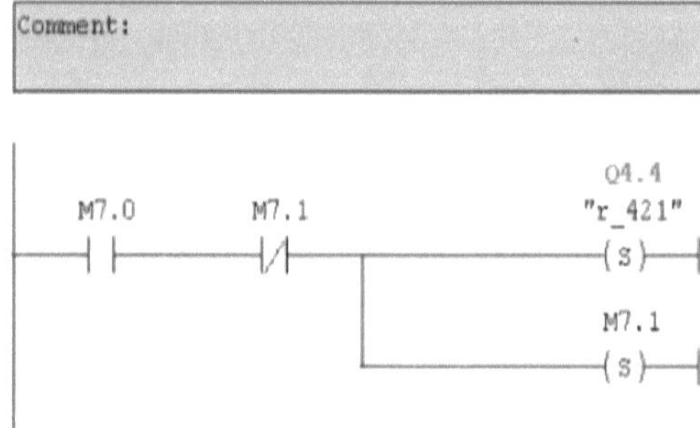

Figure 2.69

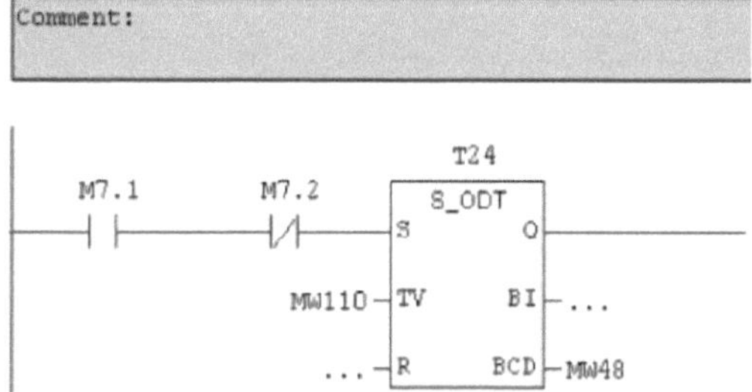

Figure 2.70

Figure 2.71

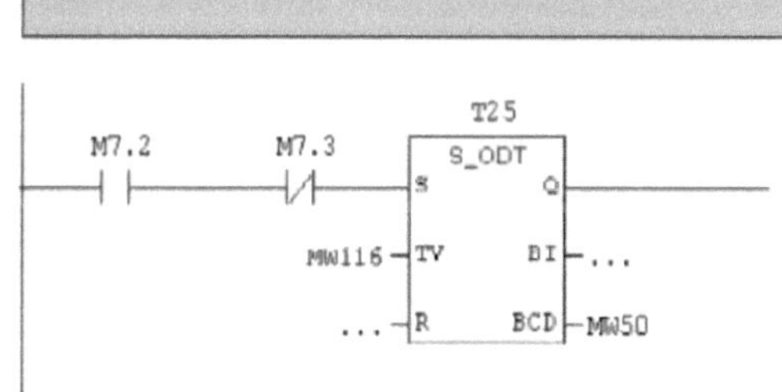

Figure 2.72

Network 61 : Title:

Comment:

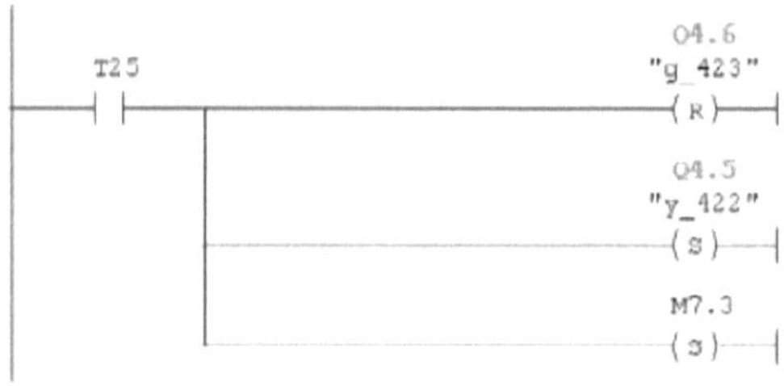

Figure 2.73

Network 62 : Title:

Comment:

```
A    M     7.3
FR   T    26
L    S5T#5S
SD   T    26
```

Figure 2.74

Network 63 : Title:

Comment:

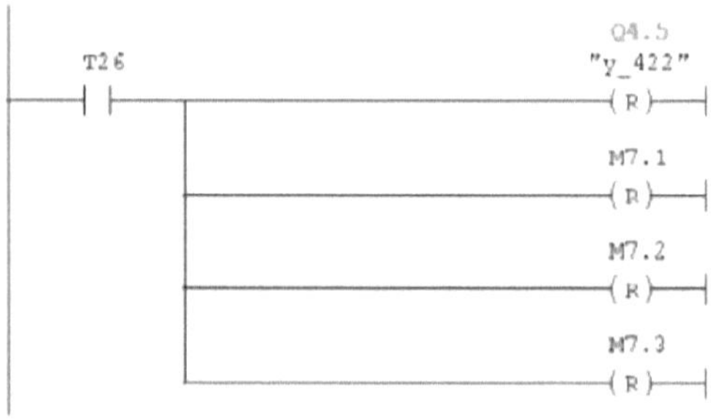

Figure 2.75

Phase 9 – The Emergency mode

As it was mentioned earlier, at 24:00 PM, system goes into its **Emergency** mode automatically. From 24:00 PM to 7:00 AM, all the yellow traffic lights on **S/N Main** road start blinking in yellow and that of **E/W**

Streets in red. And as it was mentioned in case of depressing the **Emergency** pushbutton, all flags related to the **Emergency** case, are turned on and all other flags, are reset.

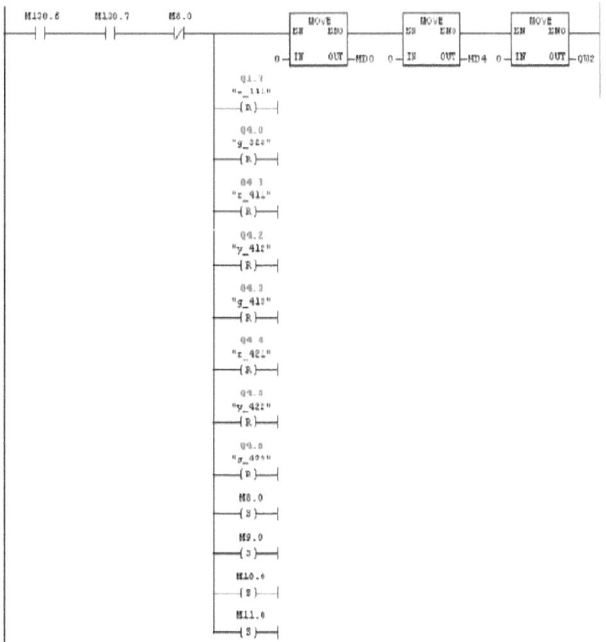

Figure 2.76

Coding to get the first intersection's traffic lights start blinking

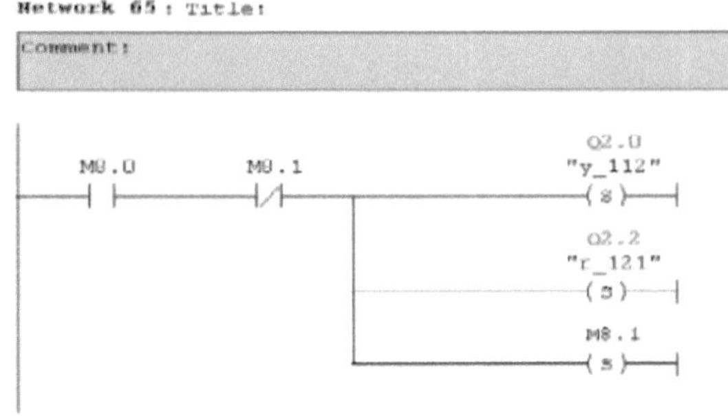

Figure 2.77

Network 66 : Title:

Comment:

```
        A     M       8.1
        AN    M       8.2

        FR    T       27
        L     S5T#500MS
        SD    T       27
```

Figure 2.78

Network 67 : Title:

Comment:

Figure 2.79

Network 68 : Title:

Comment:

```
        A     M       8.2

        FR    T       28
        L     S5T#500MS
        SD    T       28
```

Figure 2.80

Network 69 : Title:

Comment:

Figure 2.81

Coding to get the second intersection's traffic lights start blinking

Figure 2.82

Network 71 : Title:

Comment:

```
A     M     9.1
AN    M     9.2

FR    T     29
L     S5T#500MS
SD    T     29
```

Figure 2.83

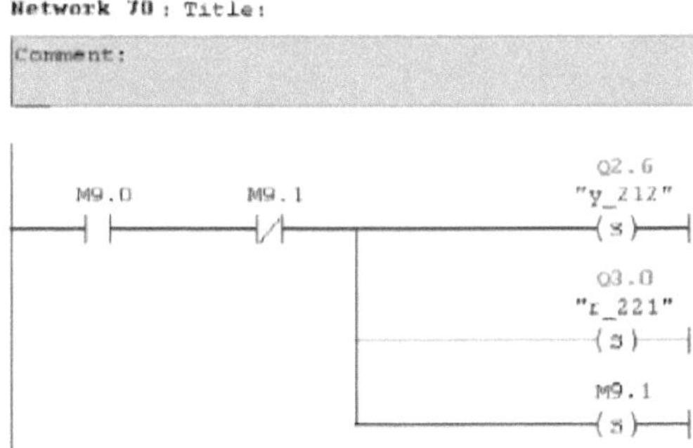

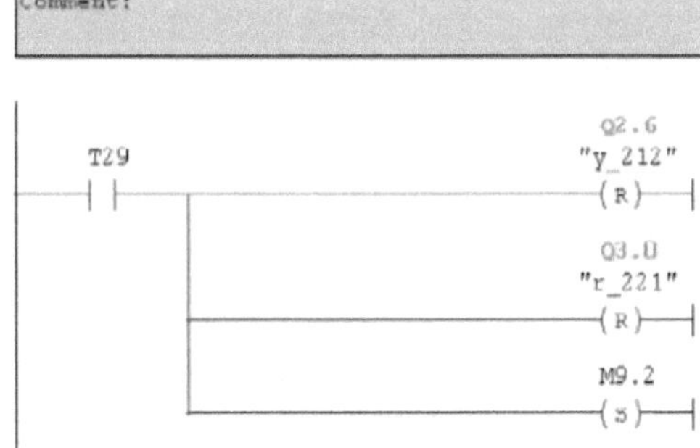

Figure 2.84

Network 73 : Title:

Comment:

```
A     M     9.2

FR    T     30
L     S5T#500MS
SD    T     30
```

Figure 2.85

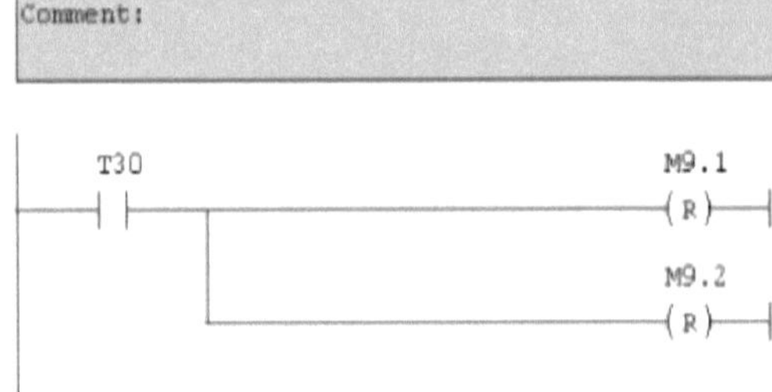

Figure 2.86

Coding to get the third intersection's traffic lights start blinking

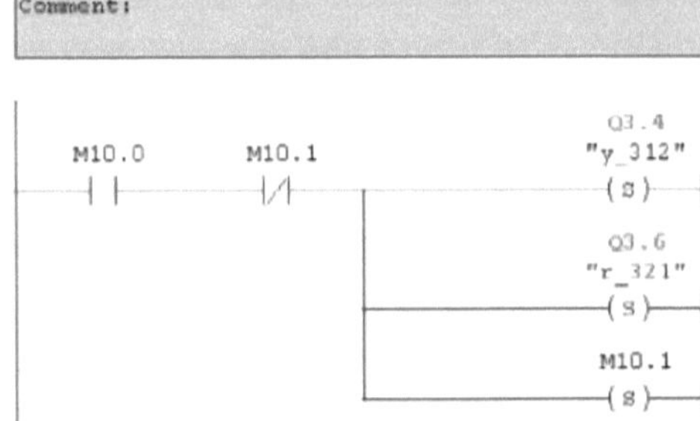

Figure 2.87

Network 76 : Title:

Comment:

```
A     M     10.1
AN    M     10.2

FR    T     31
L     S5T#500MS
SD    T     31
```

Figure 2.88

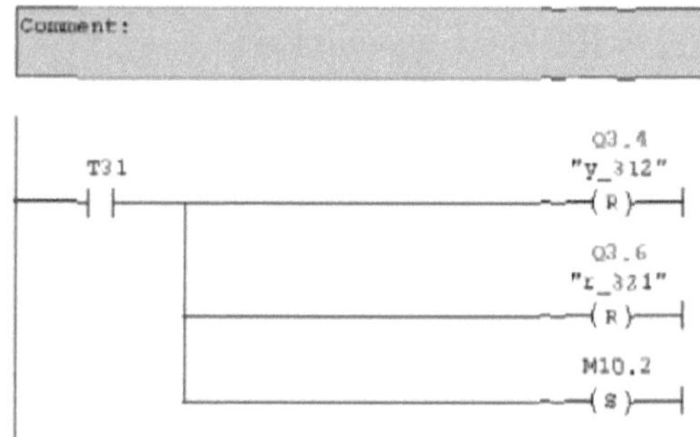

Network 77 : Title:

Comment:

Figure 2.89

Network 78 : Title:

Comment:

```
A     M     10.2

FR    T     32
L     S5T#500MS
SD    T     32
```

Figure 2.90

Figure 2.91

Coding to get the forth intersection's traffic lights start blinking

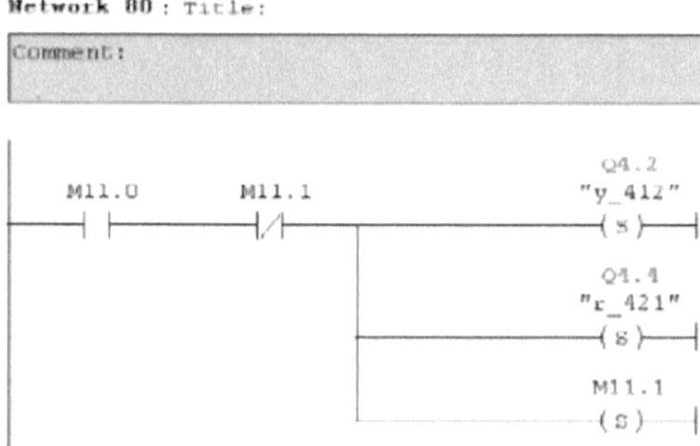

Figure 2.92

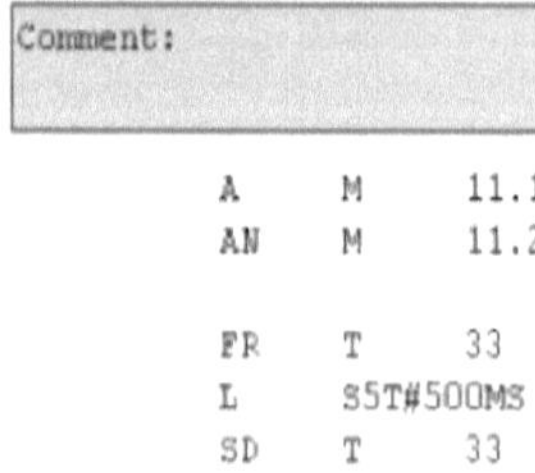

Figure 2.93

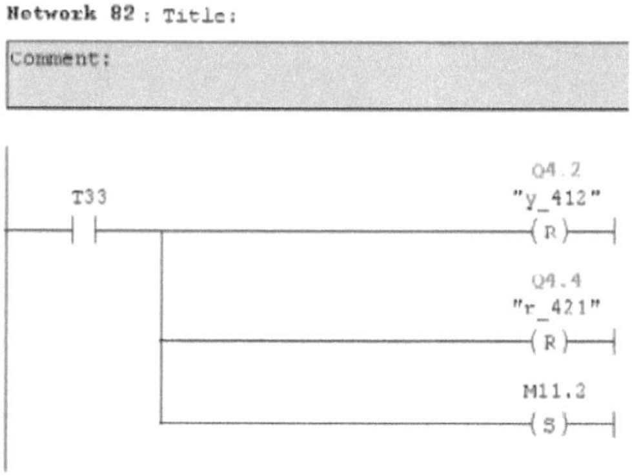

Figure 2.94

Network 83 : Title:

Comment:

```
A    M    11.2

FR   T    34
L    S5T#500MS
SD   T    34
```

Figure 2.95

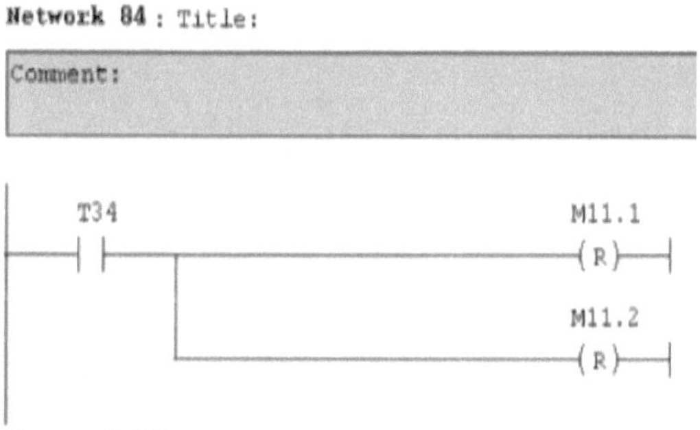

Figure 2.96

Phase 10 – Controlling the countdown timer displays

At this stage, we need to send data to all countdown timers in **BCD** format. In each intersection, we have two seven segment display PCBs. The control program should be prepared such that when either red or green LED turns on, the content of the timer which is related to that particular traffic light is sent to its display. **MOVE** instruction is used to send data to the related display PCB right at the time when condition to turn any traffic light is satisfied.

S/N side, first intersection

When M0.1 + M0.2 = 1, green light turns on, content of (MW20) > (MW52)

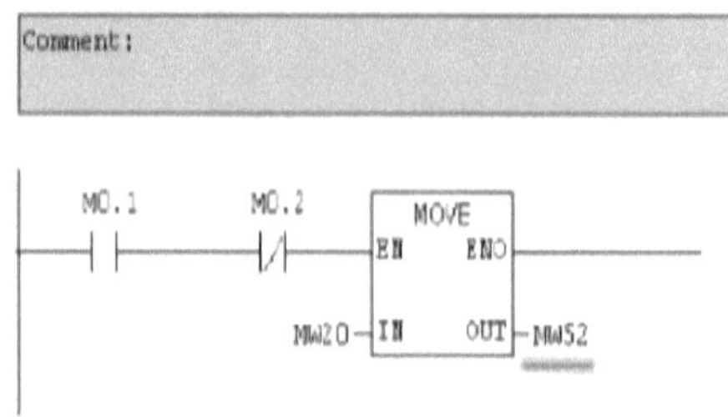

Figure 2.97

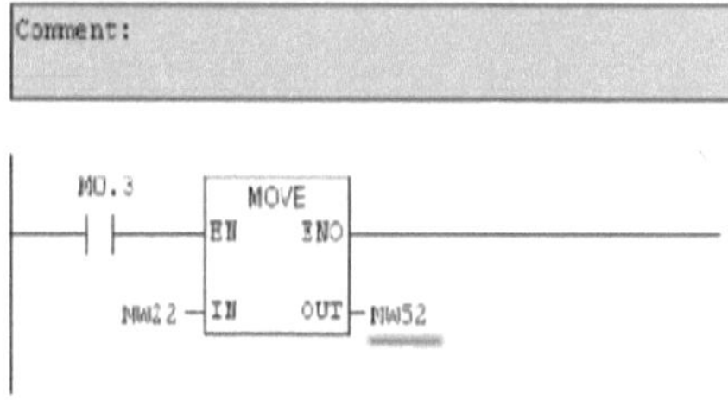

Figure 2.98

E/W side, first intersection

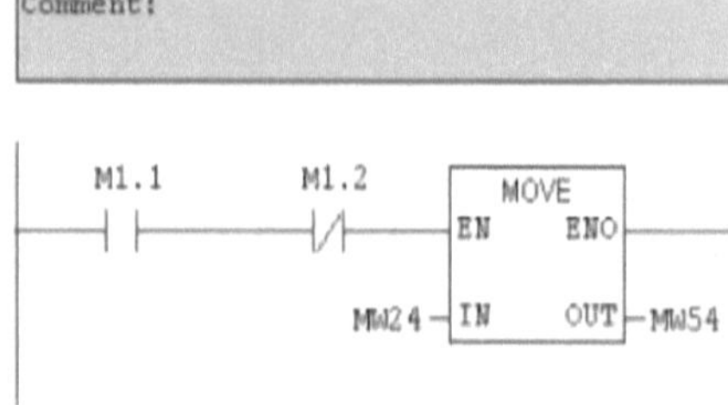

Figure 2.99

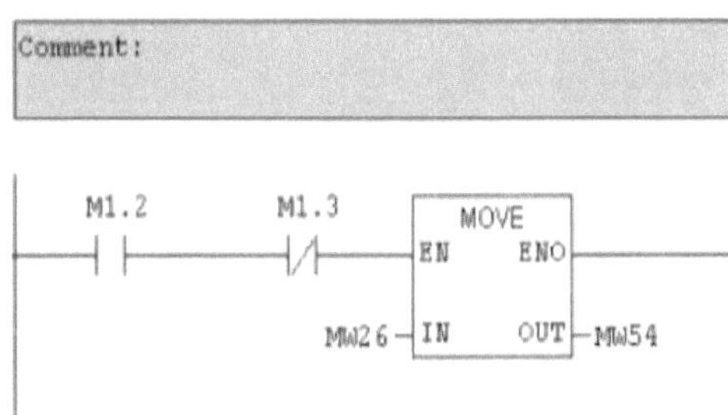

Figure 2.100

S/N side, second intersection

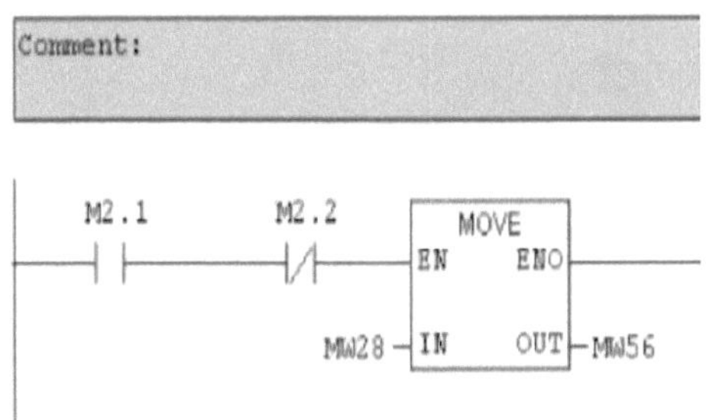

Figure 2.101

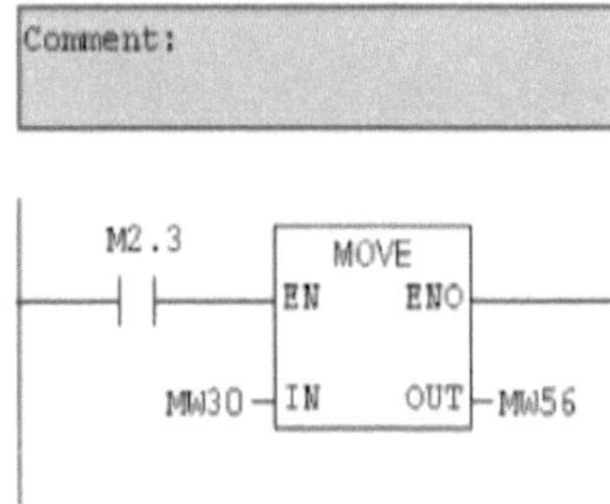

Figure 2.102

E/W side, second intersection

Figure 2.103

Figure 2.104

S/N side, third intersection

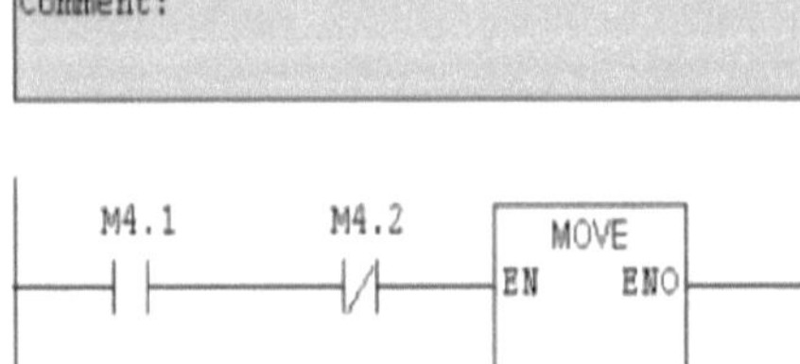

Figure 2.105

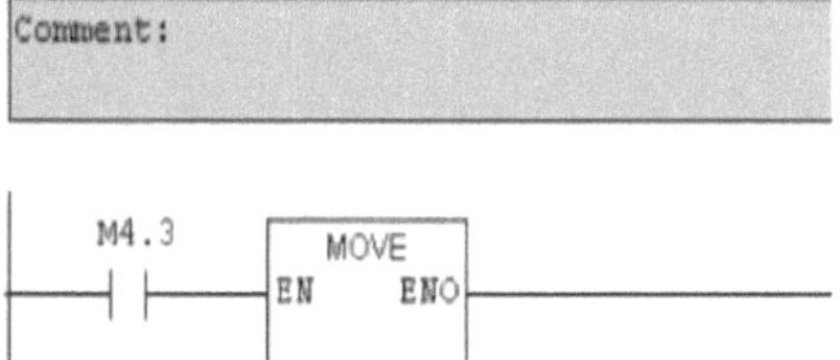

Figure 2.106

E/W side, third intersection

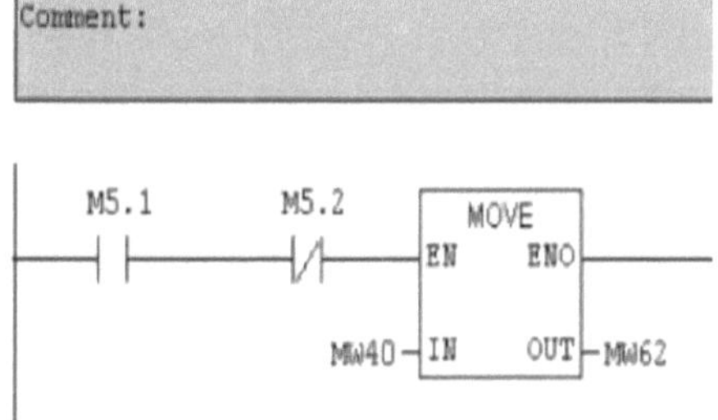

Figure 2.107

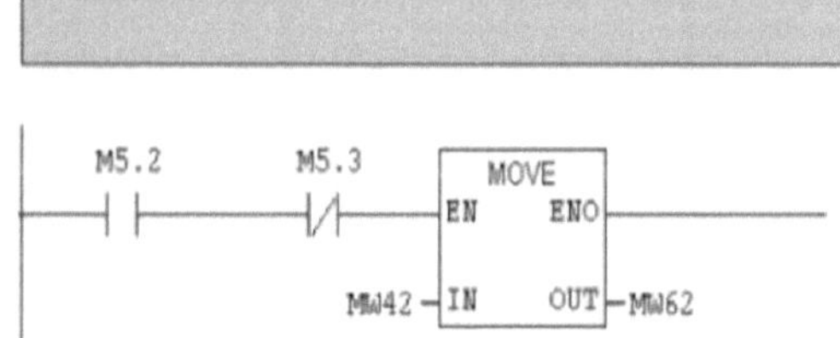

Figure 2.108

S/N side, forth intersection

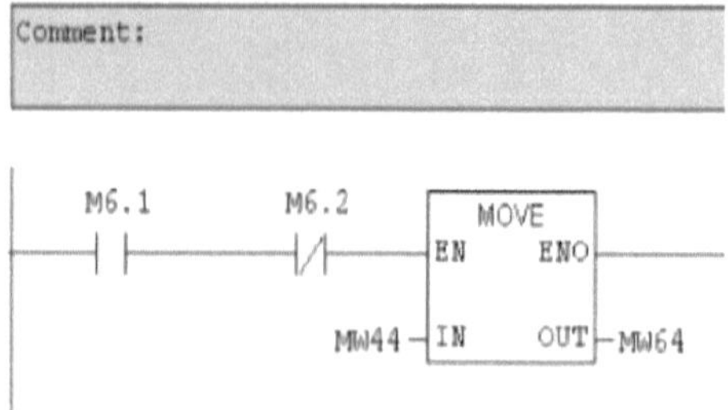

Figure 2.109

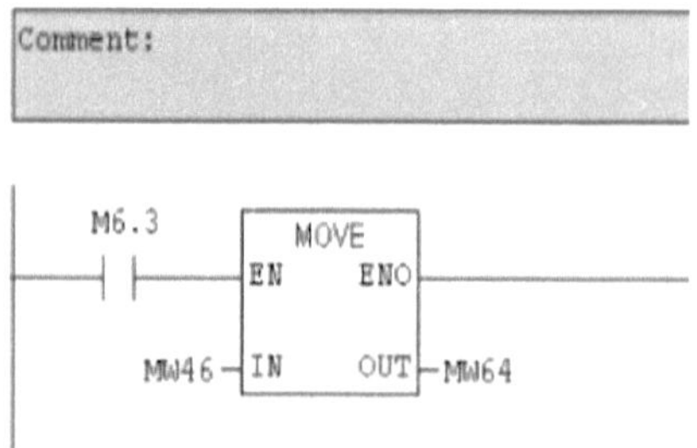

Figure 2.110

E/W side, forth intersection

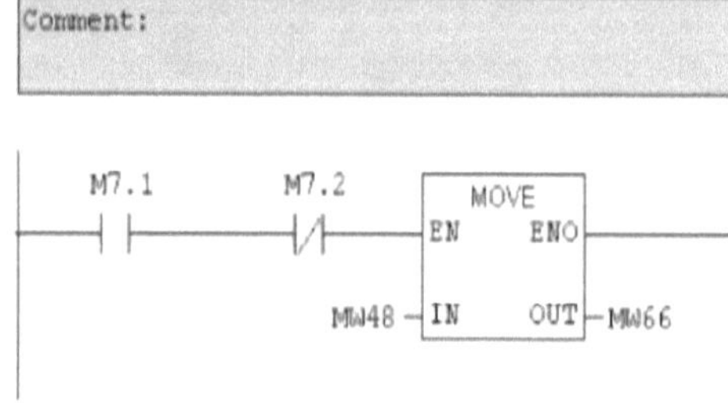

Figure 2.111

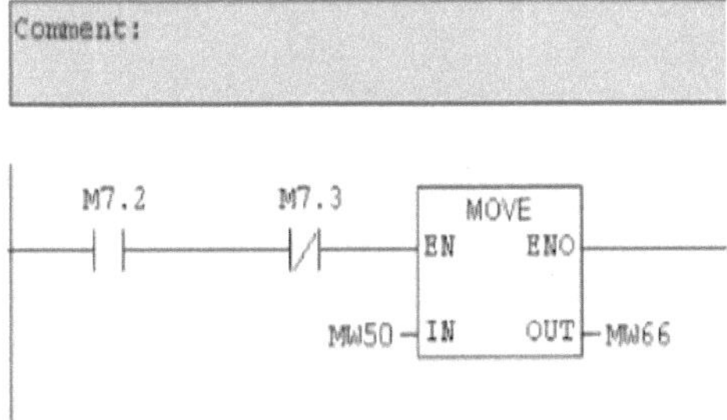

Figure 2.112

Phase 11 – Addressing the countdown timer displays

From figure 1.13, notice that 4 inputs of 2 × 74LS138 (3-line to 8 Line Decoders/ De-multiplexers) which are specified as A, B, C, and D are wired to output terminals of PLC as Q1.0, Q1.1, Q1.2, and Q1.3 respectively. And also EN input is connected to Q1.4.
In this section of the control program, we must start addressing each seven segment display PCB and in the same time, send data (in form of a BCD number) to each segment. And to get data transferred, every time when the 4 address lines are set, EN input must be set to 0 (active low) for 200 milliseconds and then it set to 1.

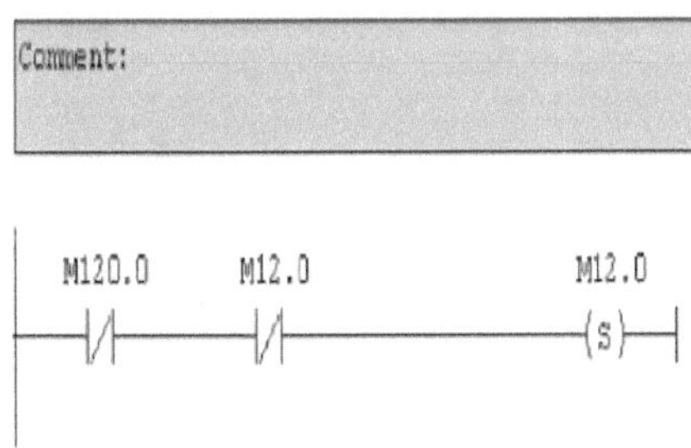

Figure 2.113

Network 103 : Title:

Comment:

```
A     M     12.0
AN    M     12.1

R     "A"                  Q1.0
R     "B"                  Q1.1
R     "C"                  Q1.2
R     "D"                  Q1.3

FR    T     35
L     S5T#100MS
SD    T     35


A     T     35
S     "A"                  Q1.0
R     "B"                  Q1.1
R     "C"                  Q1.2
R     "D"                  Q1.3

S     M     12.1
```

Figure 2.114

Next address is 0010

Network 104 : Title:

Comment:

```
A     M     12.1
AN    M     12.2

FR    T     36
L     S5T#100MS
SD    T     36


A     T     36
R     "A"                  Q1.0
S     "B"                  Q1.1
R     "C"                  Q1.2
R     "D"                  Q1.3

S     M     12.2
```

Figure 2.115

Next address is 0011

Network 105 : Title:

Comment:

```
A    M    12.2
AN   M    12.3

FR   T    37
L    S5T#100MS
SD   T    37

A    T    37
S    "A"                Q1.0
S    "B"                Q1.1
R    "C"                Q1.2
R    "D"                Q1.3

S    M    12.3
```

Figure 2.116

Next address is 0100

Network 106 : Title:

Comment:

```
A    M    12.3
AN   M    12.4

FR   T    38
L    S5T#100MS
SD   T    38

A    T    38
R    "A"                Q1.0
R    "B"                Q1.1
S    "C"                Q1.2
R    "D"                Q1.3

S    M    12.4
```

Figure 2.117

Next address is 0101

Network 107 : Title:

```
Comment:

      A     M     12.4
      AN    M     12.5

      FR    T     39
      L     S5T#100MS
      SD    T     39

      A     T     39
      S     "A"                    Q1.0
      R     "B"                    Q1.1
      S     "C"                    Q1.2
      R     "D"                    Q1.3

      S     M     12.5
```

Figure 2.118

Next address is 0110

Network 108 : Title:

```
Comment:

      A     M     12.5
      AN    M     12.6
      FR    T     40
      L     S5T#100MS
      SD    T     40

      A     T     40
      R     "A"                    Q1.0
      S     "B"                    Q1.1
      S     "C"                    Q1.2
      R     "D"                    Q1.3

      S     M     12.6
```

Figure 2.119

Next address is 0111

Network 109 : Title:

Comment:

```
A     M     12.6
AN    M     12.7

FR    T     41
L     S5T#100MS
SD    T     41

A     T     41
S     "A"                   Q1.0
S     "B"                   Q1.1
S     "C"                   Q1.2
R     "D"                   Q1.3

S     M     12.7
```

Figure 2.120

Next address is 1000

Network 110 : Title:

Comment:

```
A     M     12.7
AN    M     13.0

FR    T     42
L     S5T#100MS
SD    T     42

A     T     42
R     "A"                   Q1.0
R     "B"                   Q1.1
R     "C"                   Q1.2
S     "D"                   Q1.3

S     M     13.0
```

Figure 2.121

Next address is 1101

Network 111 : Title:

Comment:

```
A     M     13.0
AN    M     13.1

FR    T     43
L     S5T#100MS
SD    T     43

A     T     43
S     "A"                  Q1.0
R     "B"                  Q1.1
S     "C"                  Q1.2
S     "D"                  Q1.3

S     M     13.1
```

Figure 2.122

Next address is 1110

Network 112 : Title:

Comment:

```
A     M     13.1
AN    M     13.2

FR    T     44
L     S5T#100MS
SD    T     44

A     T     44
R     "A"                  Q1.0
S     "B"                  Q1.1
S     "C"                  Q1.2
S     "D"                  Q1.3

S     M     13.2
```

Figure 2.123

Next address is 1111

Network 113 : Title:

Comment:

```
A     M     13.2
AN    M     13.3

FR    T     45
L     S5T#100MS
SD    T     45

A     T     45
S     "A"                    Q1.0
S     "B"                    Q1.1
S     "C"                    Q1.2
S     "D"                    Q1.3

S     M     13.3
```

Figure 2.124

Network 114 : Title:

Comment:

```
A     M     13.3

FR    T     46
L     S5T#100MS
SD    T     46
```

Figure 2.125

At the end of this cycle, we are going to **RESET** all of the flags related to this phase. Instructions at this phase are going to be executed continuously since they are located inside of a program loop.

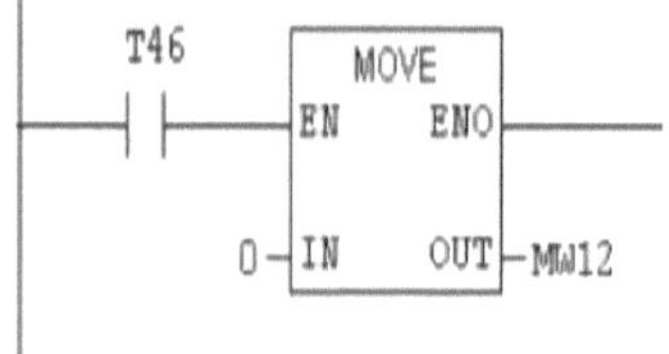

Figure 2.126

Phase 12 – Sending data to the countdown timer displays

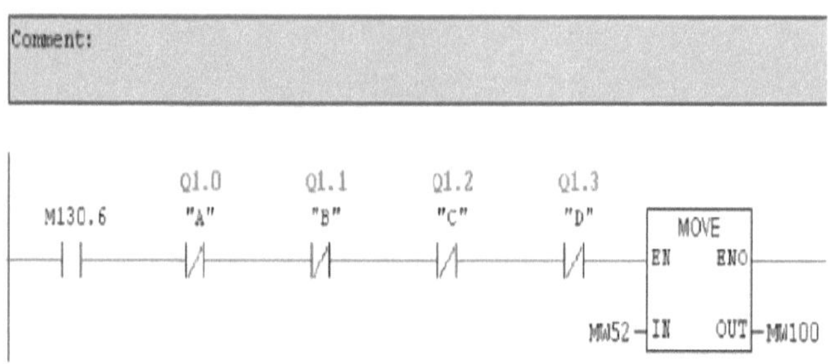

Figure 2.127

As you notice, two conditions must be satisfied to send data to each display PCB. 1- The START pushbutton flag must be set. 2- The address related to that particular display PCB must also be set as well.

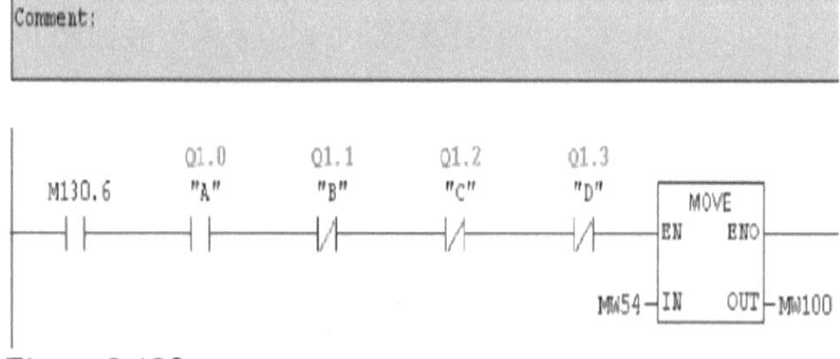

Figure 2.128

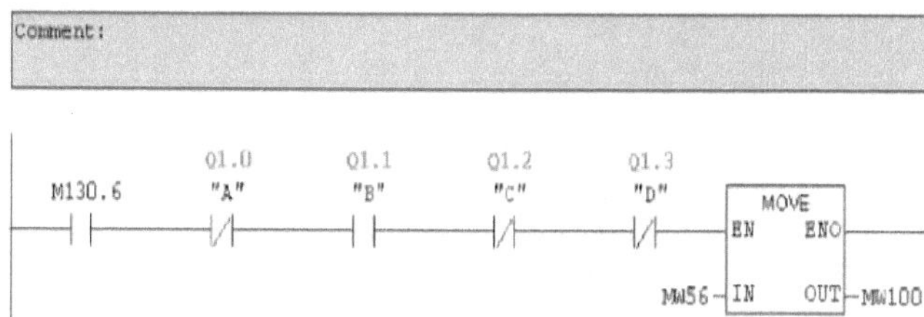

Figure 2.129

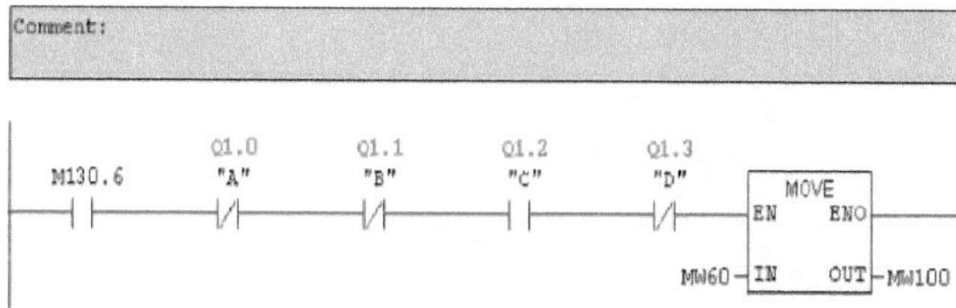

Figure 2.130

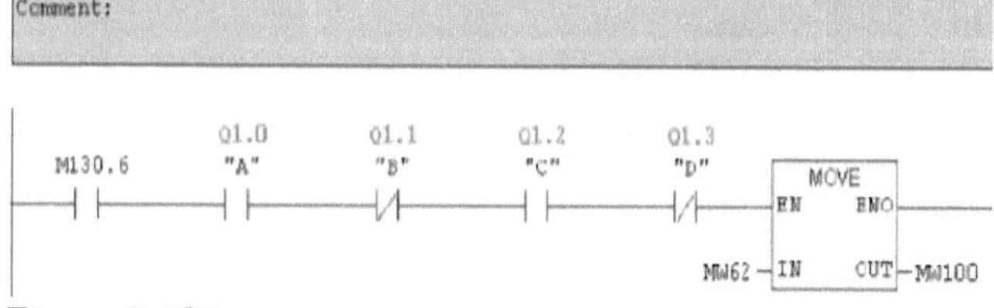

Figure 2.131

Figure 2.132

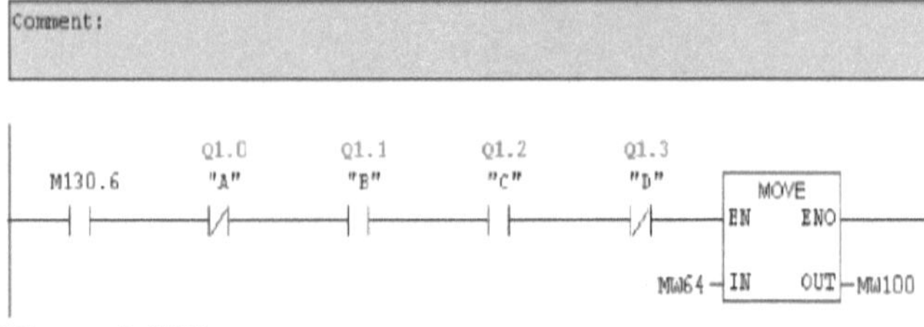

Figure 2.133

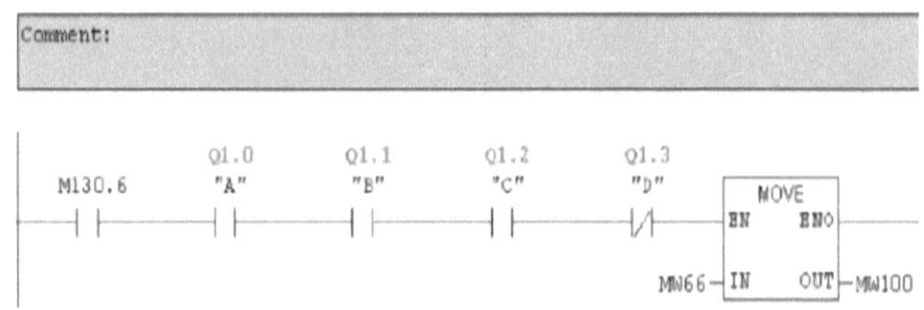

Figure 2.134

Phase 13 – Coding to get address latched by enabling the EN input

On this phase, we need to generate coding to get data latched and displayed on each display PCB. According to figure 2.135A, three steps are needed to get that done. Third step is to create an active LOW signal for 200 milliseconds. This is what we are going to do now.

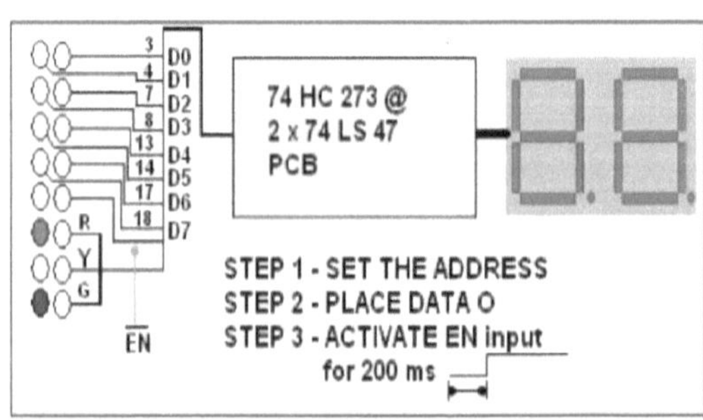

Figure 2.135A

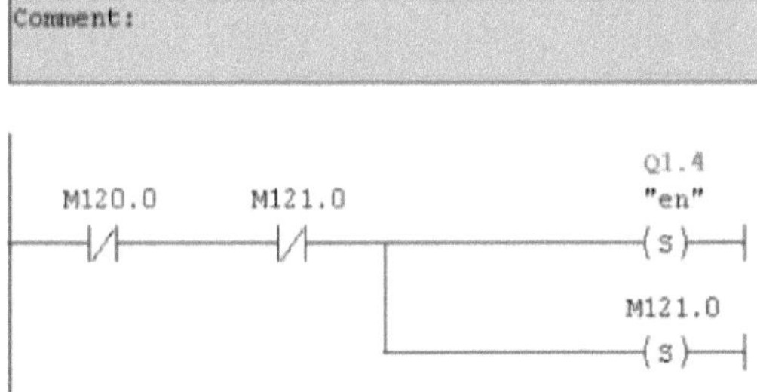

Network 116 : Title:

Comment:

Figure 2.135

Network 117 : Title:

Comment:

```
A     M     121.0
AN    M     121.1

FR    T     65
L     S5T#70MS
SD    T     65
```

Figure 2.136

Network 118 : Title:

Comment:

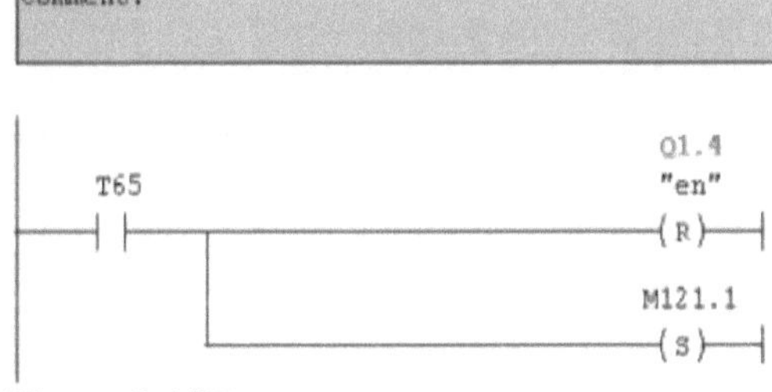

Figure 2.137

Figure 2.138

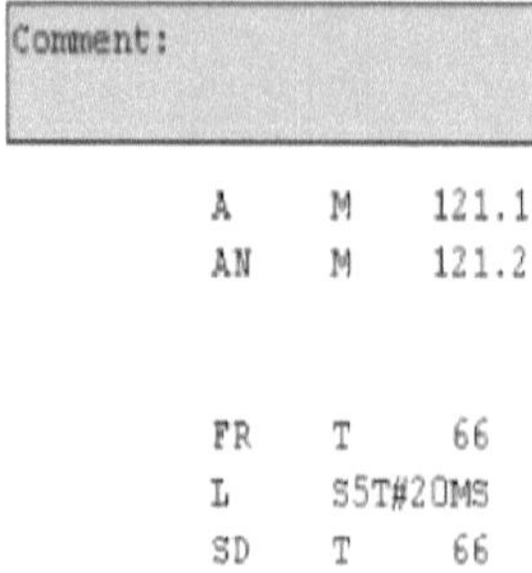

Figure 2.139

Figure 2.140

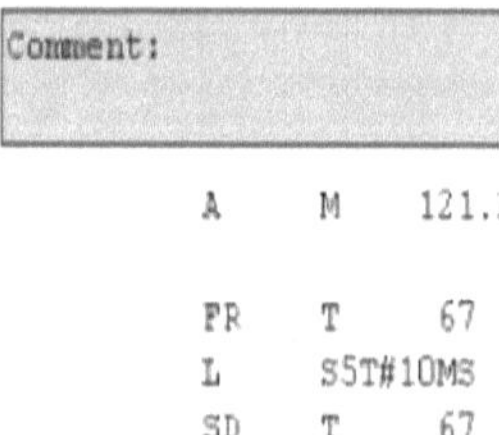

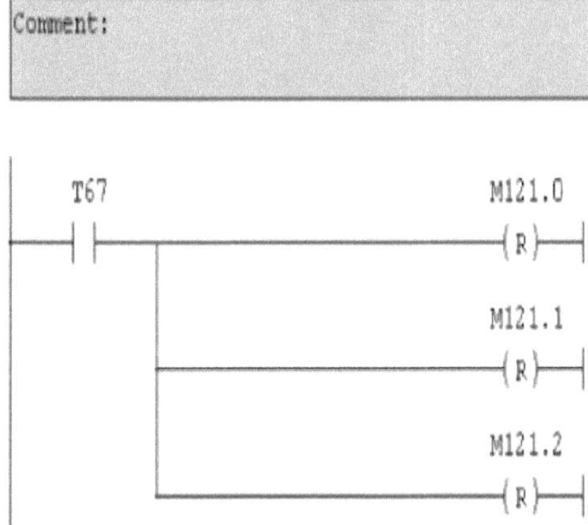

Figure 2.141

Phase 14 – Displaying number 00 on all display PCBs when system is on its emergency mode

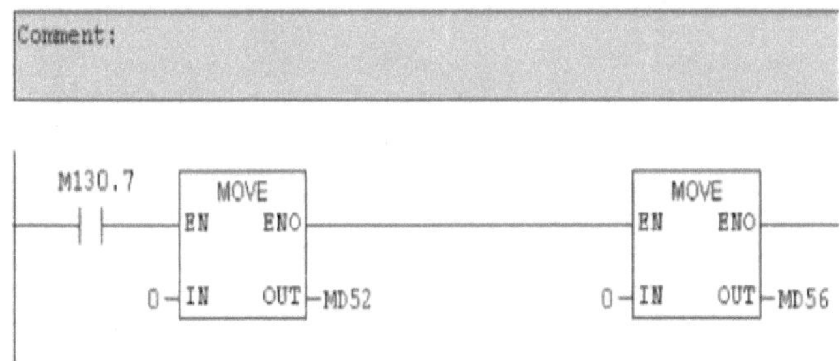

Figure 2.142

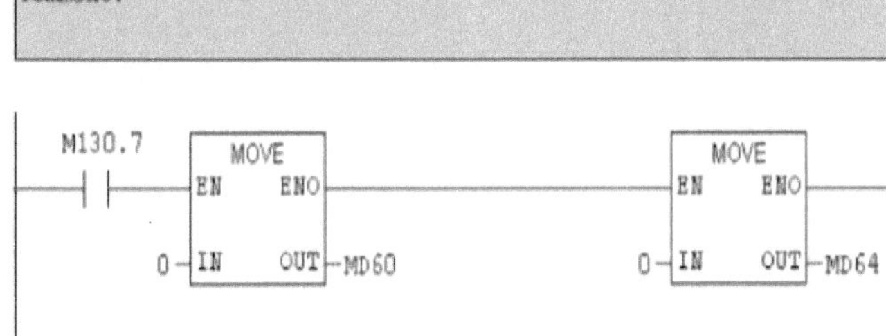

Figure 2.143

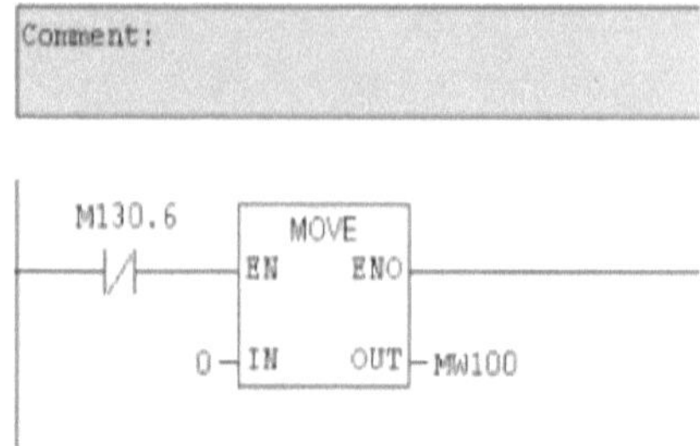

Figure 2.144

5 seconds after the STOP pushbutton is depressed, we need to make sure that all other functions of the control program is RESET and transfer of data to display PCB is stopped.

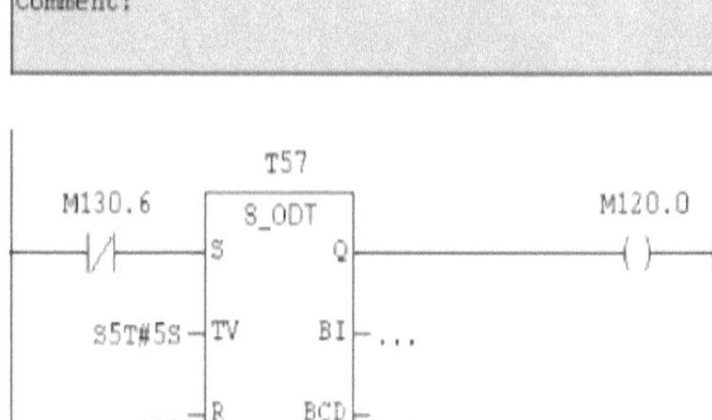

Figure 2.145

And number 00 is going to be sent on all the seven segment display PCBs.

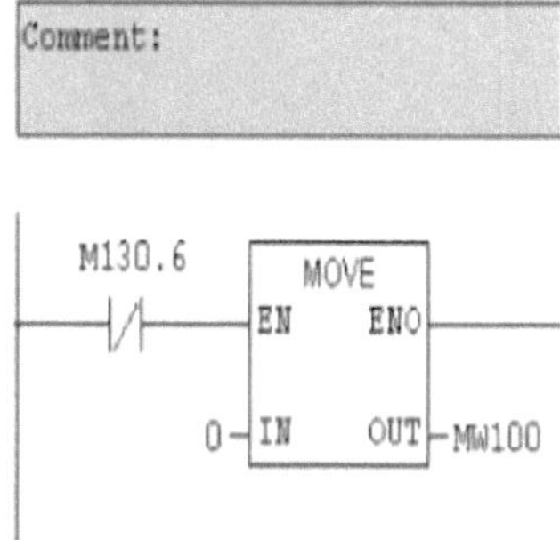

Figure 2.146

Phase 15 – Using System Function Blocks to display current time and date on the related 6 digit 7-segment display PCB.

To show current time and date, we can use SFB > SFC1 to get the job done. See figure 2.147A and 147B.

Figure 2.147A

:   :

- SFB61 RCV_PTP TEC_FUNC
- SFB62 RES_RCVB TEC_FUNC
- SFB63 SEND_RK TEC_FUNC
- SFB64 FETCH_RK TEC_FUNC
- SFB65 SERVE_RK TEC_FUNC
- SFB75 SALRM DP
- SFC0 SET_CLK CLK_FUNC
- SFC1 READ_CLK CLK_FUNC
- SFC2 SET_RTM CLK_FUNC
- SFC3 CTRL_RTM CLK_FUNC
- SFC4 READ_RTM CLK_FUNC

Figure 2.147B

Figure 2.148

The output of this function is a 64 bit number that is saved as content of the
TEMP variable as **date and time** data.

Name	Data Type	Address	Comment
OB1_PRIORITY	Byte	2.0	Priority of OB Execution
OB1_OB_NUMBR	Byte	3.0	1 (Organization block 1, OB1)
OB1_RESERVED_1	Byte	4.0	Reserved for system
OB1_RESERVED_2	Byte	5.0	Reserved for system
OB1_PREV_CYCLE	Int	6.0	Cycle time of previous OB1 scan (milliseconds)
OB1_MIN_CYCLE	Int	8.0	Minimum cycle time of OB1 (milliseconds)
OB1_MAX_CYCLE	Int	10.0	Maximum cycle time of OB1 (milliseconds)
OB1_DATE_TIME	Date_And_Time	12.0	Date and time OB1 started
my_time	Date_And_Time	20.0	

Figure 2.149

Figure 2.149A displays the content of the MWs related to **time and date** of the system.

MW xx	content
MW 68	YEAR
MW 72	MONTH
MW 70	DAY
MW 78	HOUR
MW 76	MINUTE
MW 74	SECOND

Figure 2.149A

Now need to generate coding to show the current time and date. Since we have only one display PCB to use, program is going to switch between showing **Time** for one minute and switch to **Date** at the beginning of next minute. We are going to use a flag that is SET for one minute and is **RESET** for one minute.

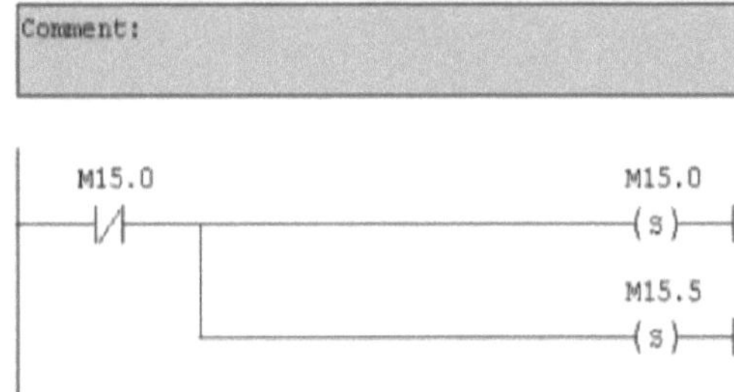

Figure 2.150

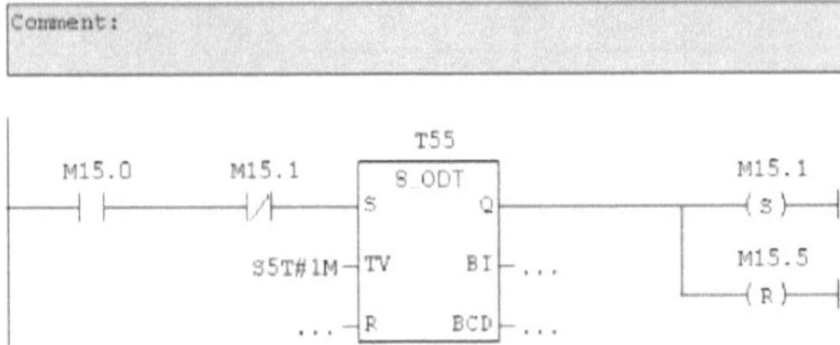

Figure 2.151

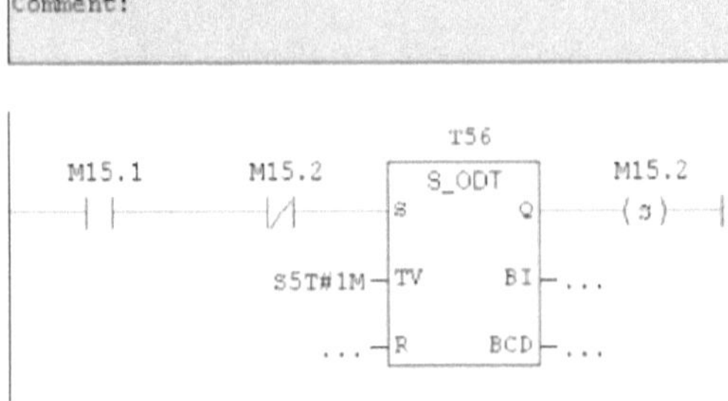

Figure 2.152

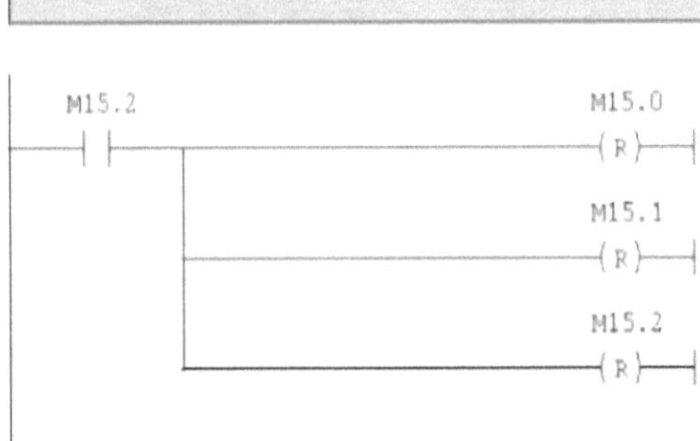

Figure 2.153

Generating codes to send time and date info on to the related seven segment PCBs.

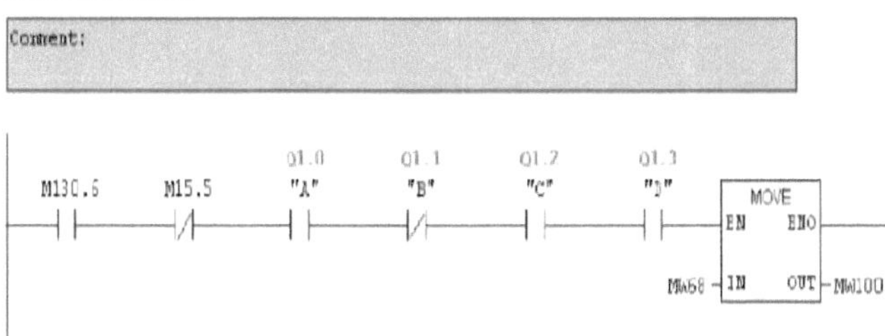

Figure 2.154

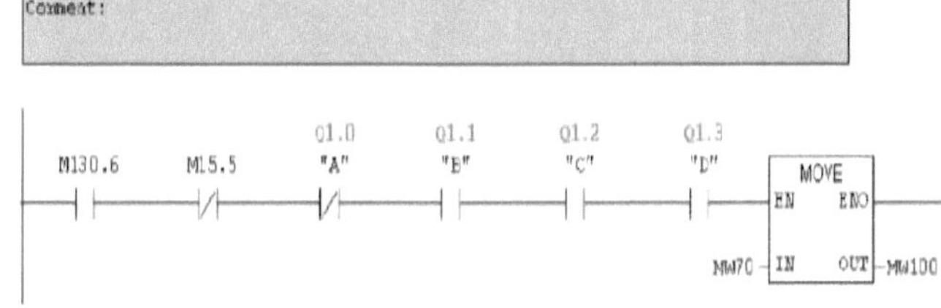

Figure 2.155

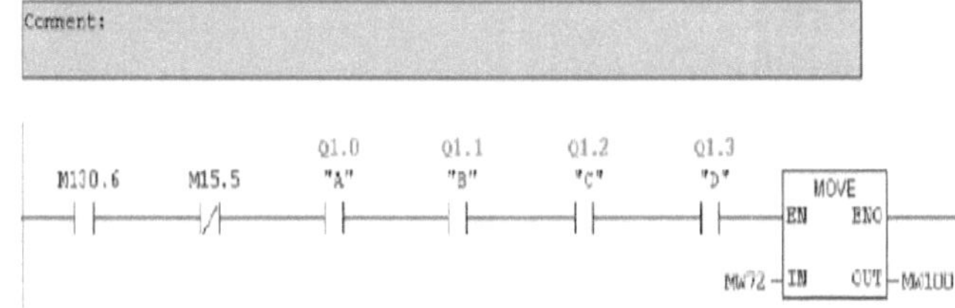

Figure 2.156

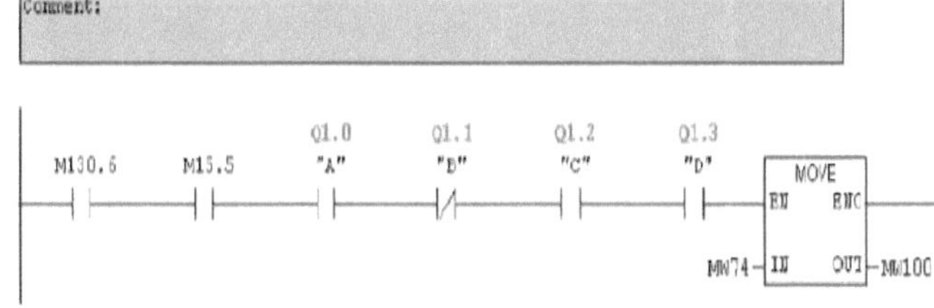

Figure 2.157

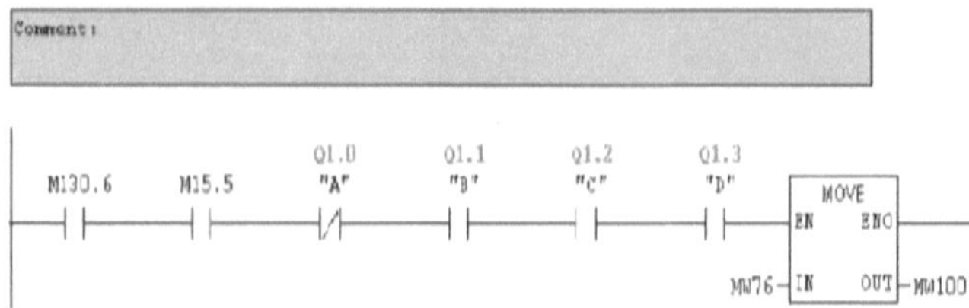

Figure 2.158

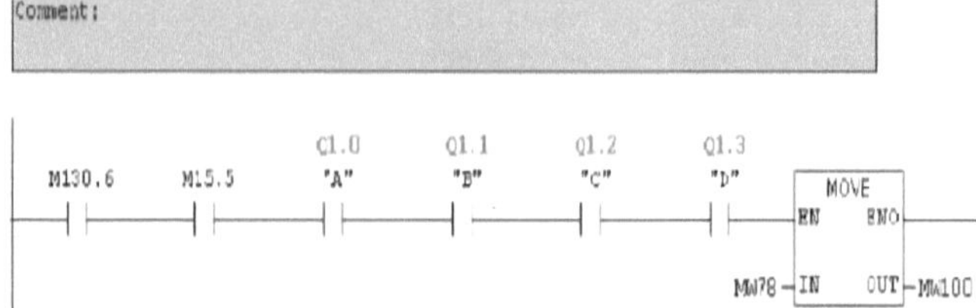

Figure 2.159

When the Time and Date PCB is displaying the Time, total of 4 LEDs start blinking (2 of each located between minute and second digits).When Date is displayed, only 2 LEDs start blinking. See figure 1.4. The following codes get the LEDs to blink at the right time.

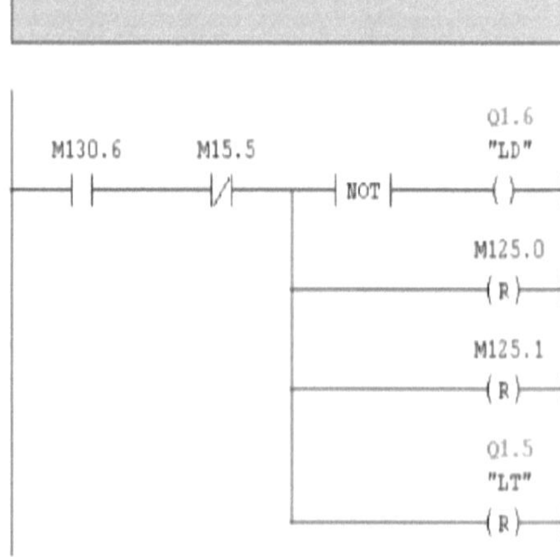

Figure 2.160

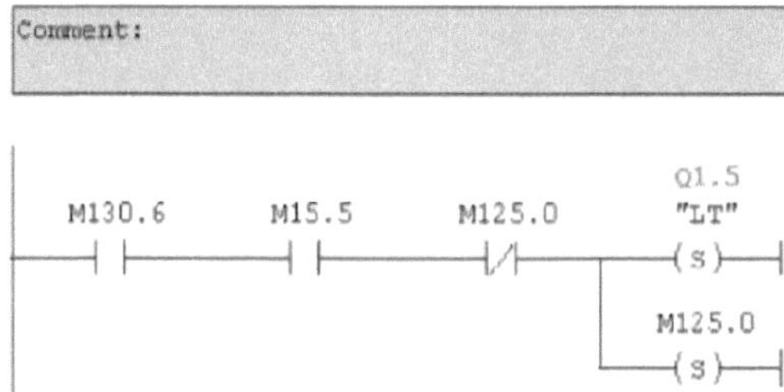

Figure 2.161

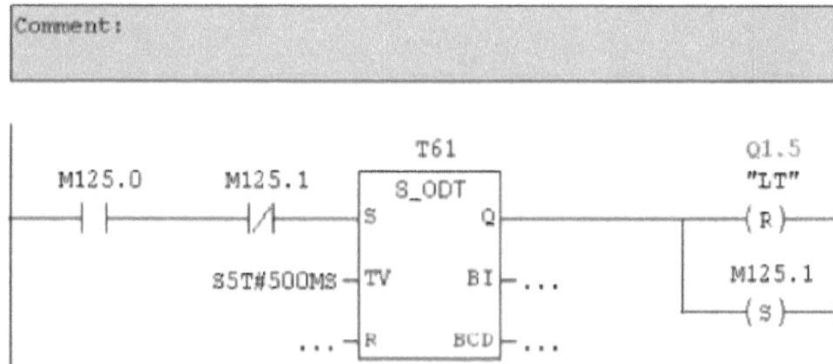

Figure 2.162

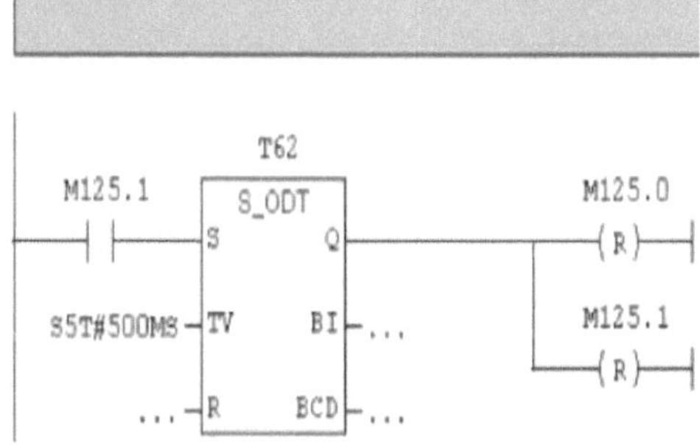

Figure 2.163

Phase 16 – Measuring speed of a passing car and comparing it with 30 Km/h speed limits

To measure speed of a passing car, we need to SET a flag and turn on a Timer when any car passes sensor S1 at any time.

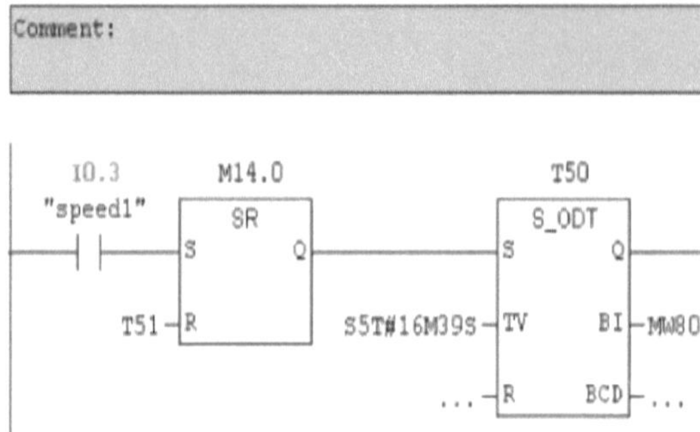

Figure 2.164

When the same car passes the second sensor S2, the amount of time elapsed to travel 100 meters (the distance between two sensors S1 and S2 is assumed 100 meters) is going to be written in a defined memory word.

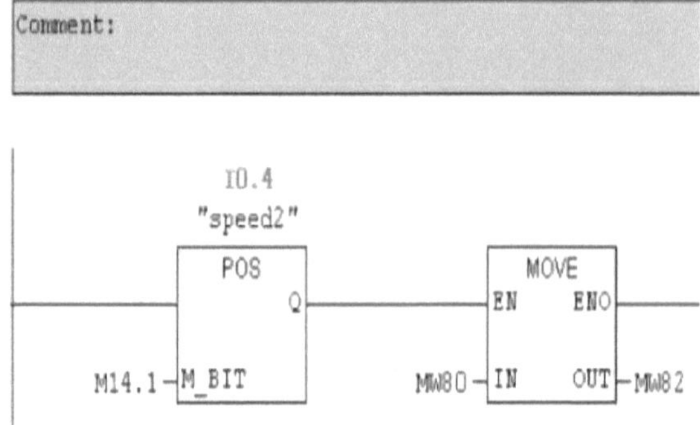

Figure 2.165

200 milliseconds, after activation of the second sensor S2, the Timer which was activated by S1 must be RESET to get it ready to measure the speed of the next car.
Check the following Codes to satisfy the last requirement.

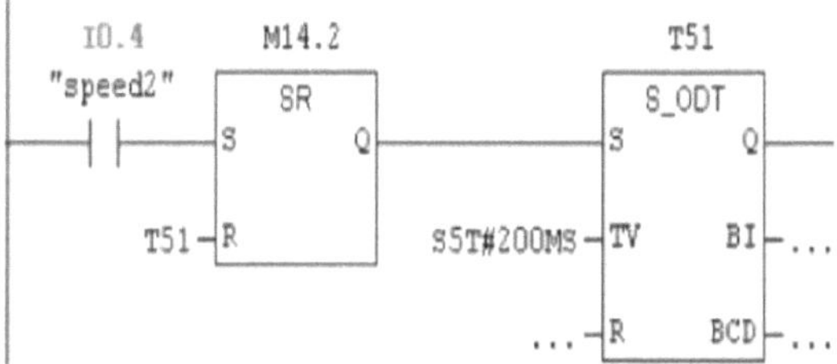

Figure 2.166

1- Since Timer T51 counts down from a set value (999) to show the final counted value, then we must deduct the final value from 999 to get the measured time value or the amount of time measured for the car to reach from first sensor (S1) to the second one (S2).

2- The equation to calculate speed of a car is $V = X \div T$. Assuming X = 100 meters, and time is in **second**, then the final equation is $V = 360 \div T$ in which **V** is in terms of **Km/h**, and **Time** in Second. Or in general we can say:

$$1 \text{ m/s} = 3.6 \text{ km/h} .$$

The distance between the first and the second intersection is 5 KM, between the second and the third one is 15 KM, between the third and the forth one is 20 KM. For a driver to travel these distances within 30 Km/h required speed limit, it will take him 10, 30, and 40 minutes respectively to receive green lights passing the traffic intersections.

5 Km = 30 Km/h × T	T= 5/30 = 1/6 H × 60 = 10 minutes
15 Km = 30 Km/h × T	T = 15/30 = 0.5 h = 0.5 × 60 = 30 minutes
20 Km = 30 Km/h × T	T = 20/30 = 2/3 h = 2/3 × 60 = 40 minutes
100 m passed @ 2 seconds	V (in Km/h) is (100 ÷ 2) × 3.6 = 180

Based on what is said about calculating the speed of a car, we generate and add codes for the PLC to calculate the speed and display it on the 2-digit display PCB.

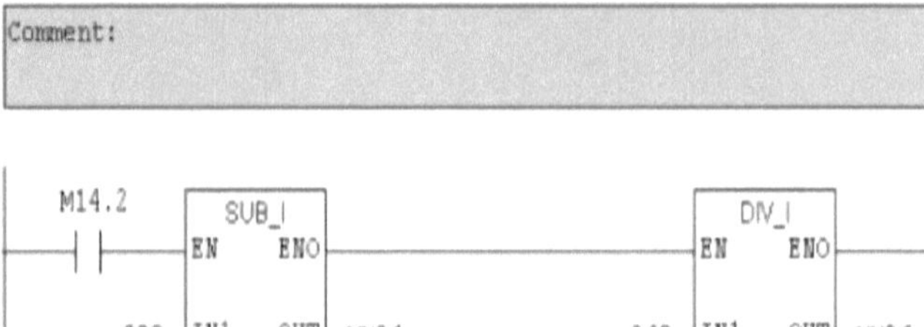

Figure 2.167

On this phase, the speed is calculated. Now need is to display it for 20 seconds and if there is not any other car passing, the speed value to be cleared from display PCB and 00 to be shown instead of the speed value.

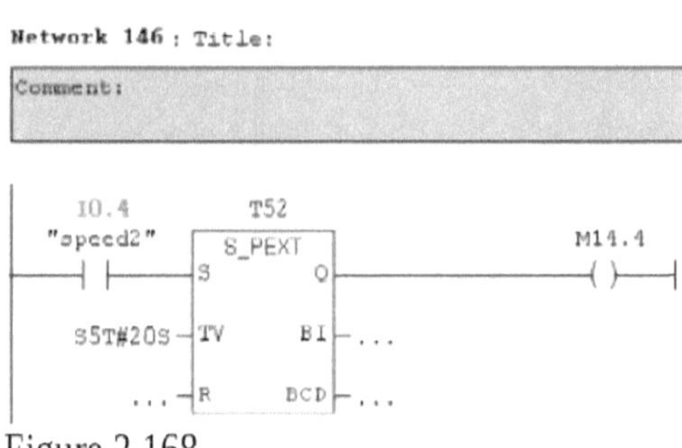

Figure 2.168

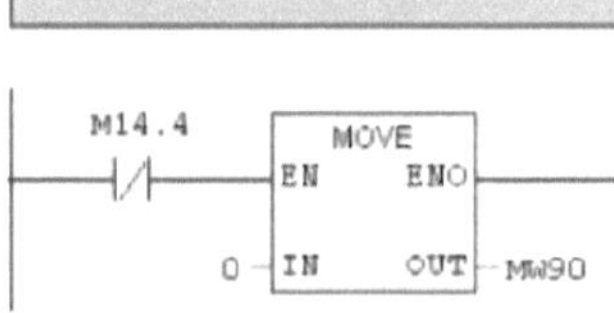

Figure 2.169

Now it is time to compare the calculated speed with 30 Km/h speed limits. If it is greater than 30 Km/h, then red LED to be turned on otherwise, the green one to be turned on.

Also the calculated speed value, which is in form of an integer number, must be changed into a BCD number to be displayed on the 7-segment display PCB.

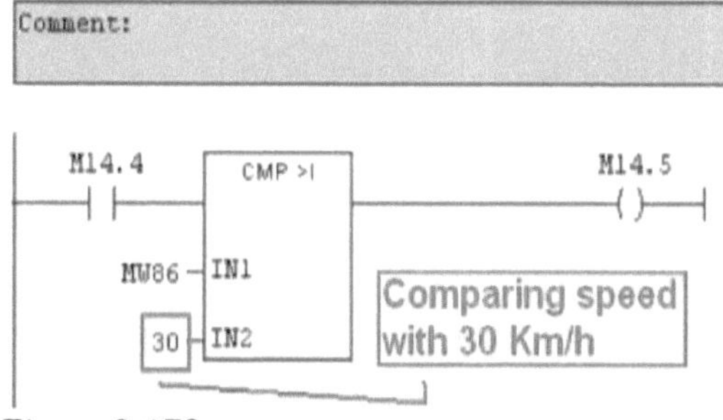

Figure 2.170

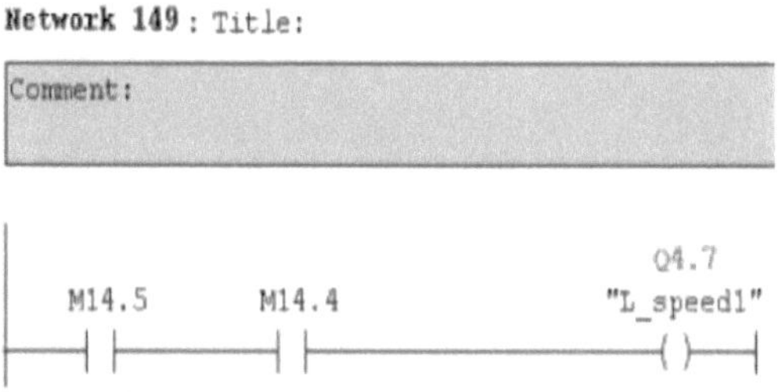

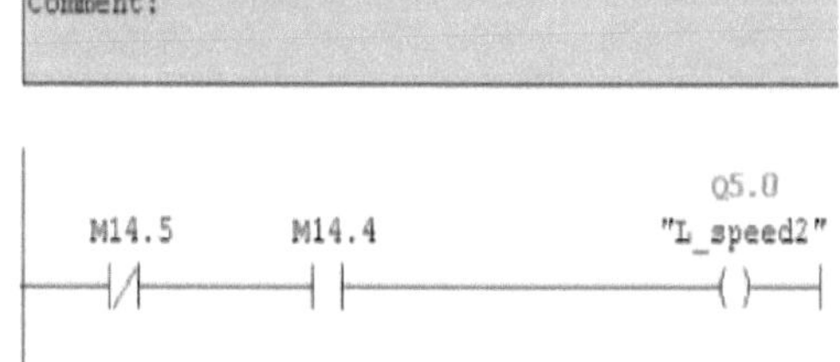

Figure 2.171

Figure 2.172

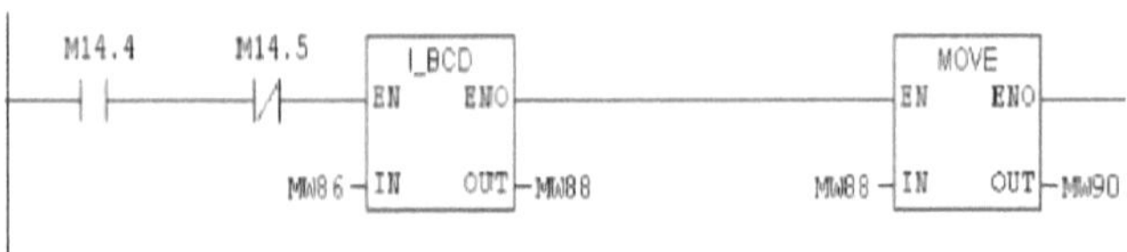

Figure 2.173

Figure 2.174

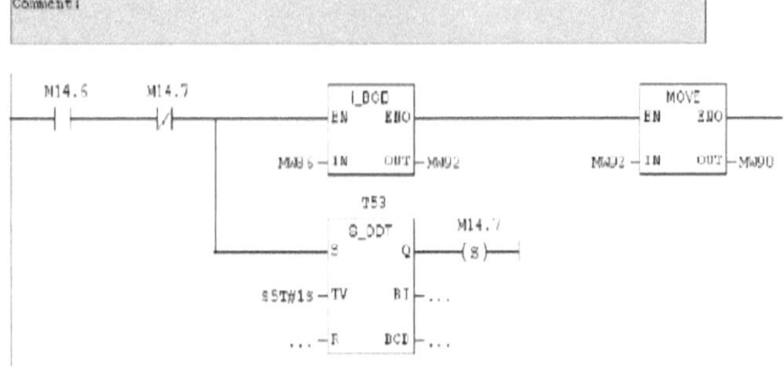

Figure 2.175

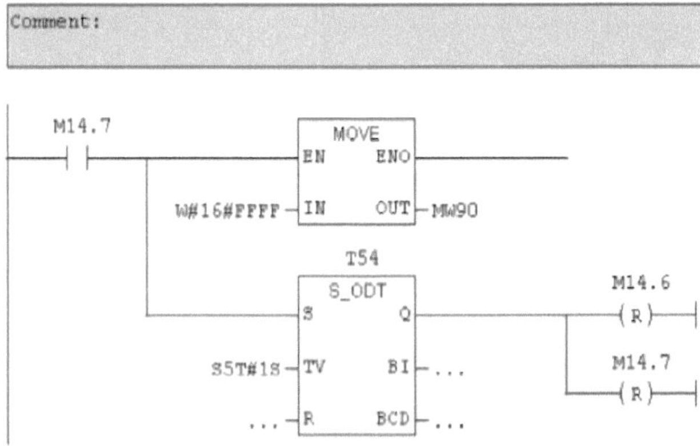

Figure 2.176

Now the speed amount value is ready to be displayed.

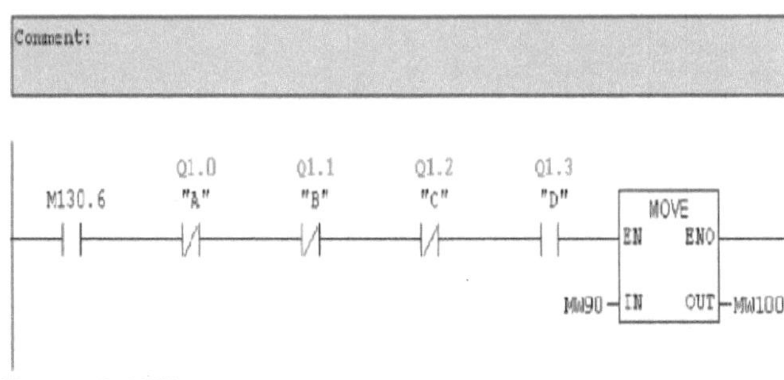

Figure 2.177

Phase 17 – The 7-segment display data which is saved at MW100, now must be sent to a related display PCB

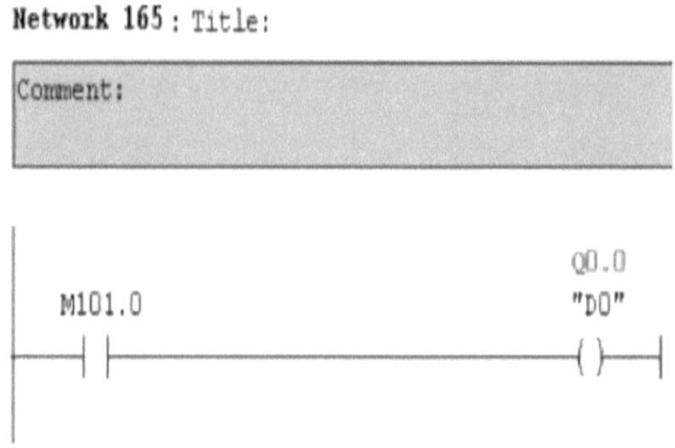

Figure 2.178

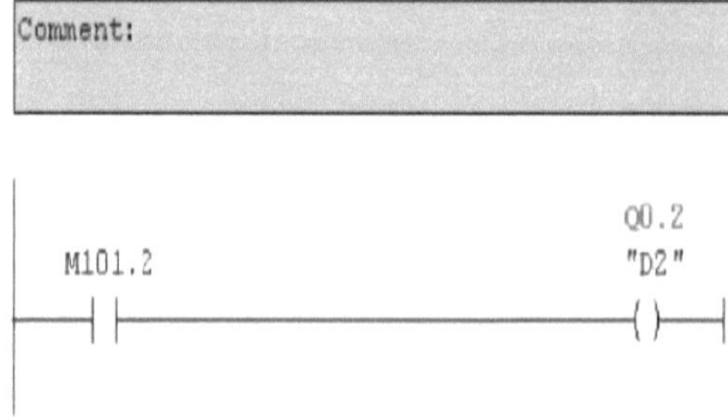

Figure 2.179

Figure 2.180

Network 168 : Title:

Comment:

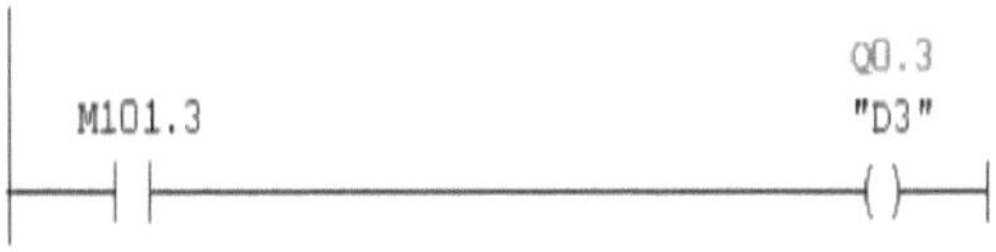

```
                                          Q0.3
     M101.3                               "D3"
----| |----------------------------------( )----
```

Figure 2.181

Network 169 : Title:

Comment:

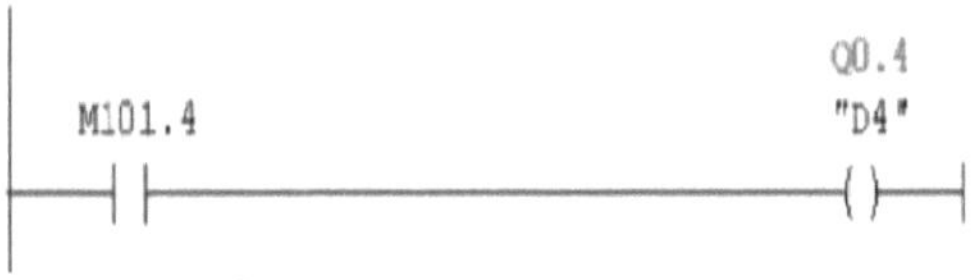

```
                                          Q0.4
     M101.4                               "D4"
----| |----------------------------------( )----
```

Figure 2.182

Network 170 : Title:

Comment:

```
                                          Q0.5
     M101.5                               "D5"
----| |----------------------------------( )----
```

Figure 2.183

Network 171 : Title:

Comment:

```
                                          Q0.6
     M101.6                               "D6"
----| |----------------------------------( )----
```

Figure 2.184

Figure 2.185

Now it is time to connect the MPI cable to the PLC and Hardware system and download the developed control program to the PLC's user memory section and see all good things happening! Remember Ch_ra.s7p file.

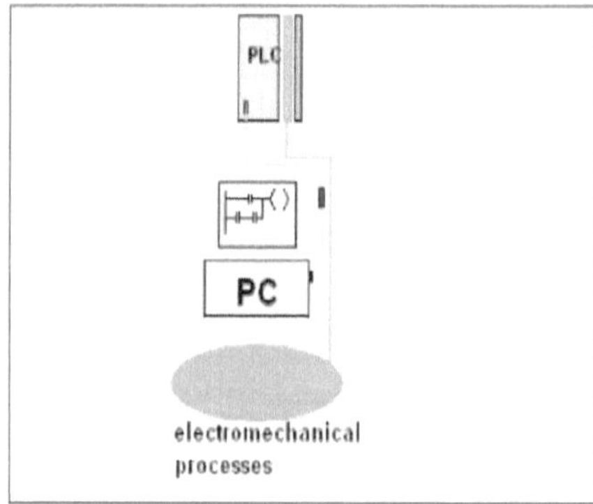

Figure 2.186

Adding an HMI display to our project

In this chapter, we are going to develop the HMI program to see all good things happening on the PCBs: depression of pushbuttons, calculation speed of the traveling car from S1 to S2 etc on a 7 inch HMI display developed by PanelMaster Company which is connected to the PLC's related port.

I am going to explain briefly how such a program can be developed for the said LCD display. And, if you are stuck up anywhere while designing the screen graphic on your own, or having difficulty to understand what I am talking about, you can always install the PM Designer software (included with this CD ROM) on your system and take a look at the settings by checking out the final program I developed.

Remark

1- You can always download free version of the PM Designer software from the following link:
www.lumel.com.pl/download/.../panel_master_designer.pdf
2-There is a tutorial document regarding the application of the PM Desinger softwre in the following link:
http://www.plc-doc.com/category/online-tutorial/simulation-and-design-software/
3- The name of the code file developed for this section of the project is v1.pm2.

<u>**Phase 18 – Installation of a typical HMI LCD touch control panel**</u>

Three steps to take before the usage of the device are:
1- Install the device on a fixture and connect it to a 24 V DC power source.
2- Connect it to a PLC via a suitable cable as seen on <u>figure 3.1</u>.
3- Use any of COM1, COM3, or USB port or your flash memory device to program the panel prior of its connection to the project PLC with a PC computer. See <u>figure 3.1</u>.

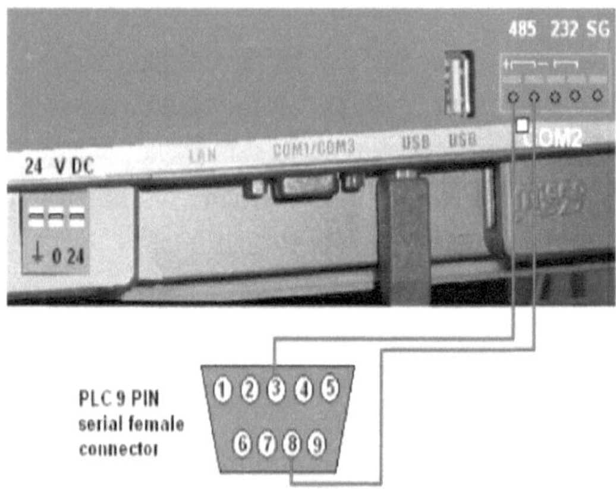

Figure 3.1

You have to program the device (from now on, the word "device" means HMI device) first, after that is done, you can connect it to the PLC (via its RS 485 port) which is supposed to execute the v1.pm2 application program.

PanelMaster Company has developed software named "**PM Designer**" which is free and can be downloaded from internet to be used to program the device. In

<u>**Phase 19 – Altering the control program to establish connection to HMI**</u>

Since we have three pushbutton symbols (START, STOP, and Emergency) to issue commands from the **HMI** device to **PLC** running the control program, then in the main control program we must add some parallel flags to simulate the related inputs to cause the same effect when any of inputs is depressed from the **device**.

[Figure 3.1A](#) displays Memory Words holding data to be written on the HMI display.

Description	MW/Flag #	Network #	Description	MW/Flag #	Network #
Countdown timer # 1	MW52	85-86	SPEED	MW90	151
2	MW54	87-88	START	m132.1	176
3	MW56	89-90	STOP	m132.2	177
4	MW58	91-92	EMERGENCY	m132.3	173
5	MW60	93-94	MONTH/hours	MW134	178-179
6	MW62	95-96	DAY/minutes	MW136	178-179
7	MW64	97-98	YEAR/second	MW138	178-179
8	MW66	99-100			

Figure 3.1A

Figure 3.2

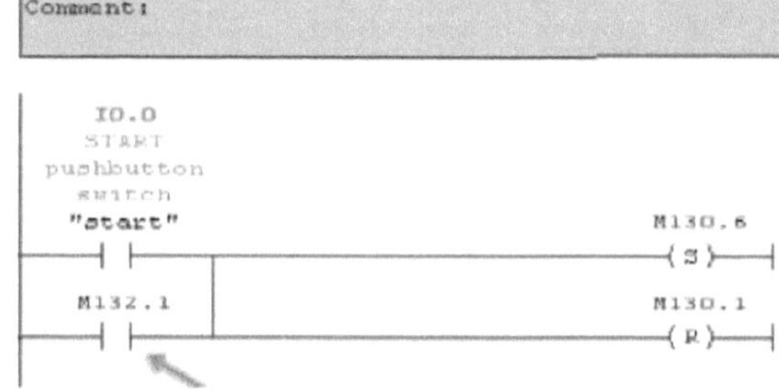

Figure 3.3

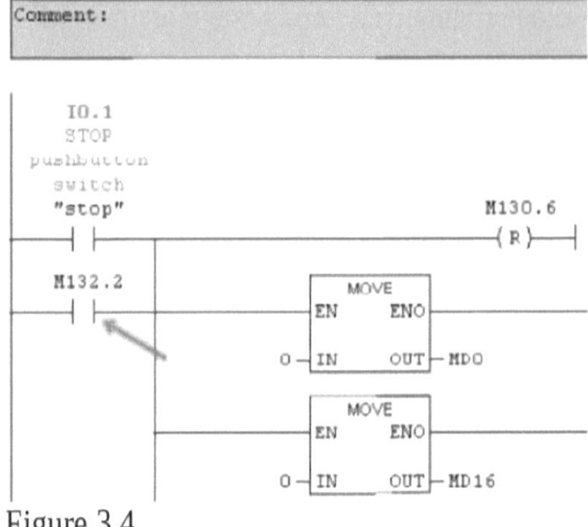

Figure 3.4

To display current time and date, we must create new memory words. Hence, we can move data related to Time and Date to the new memory words (MW) which can be used by the device. Figure 3.4A compares memory words whose contents are used to show time and date on the 7-segment PCB and the equivalent ones whose content would be used by the HMI device. In figure 3.5 notice that M132.0 turns LEDs related to the **Date** seven segment PCB. Any time it is on, the current **Date** is being displayed.

MW xx	HMI DATA	content
MW 68	138	YEAR
MW 72	134	MONTH
MW 70	136	DAY
MW 78	134	HOUR
MW 76	136	MINUTE
MW 74	138	SECOND

Figure 3.4A

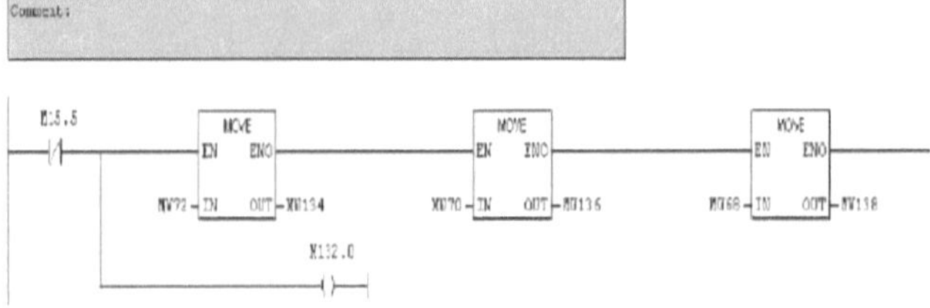

Figure 3.5

Figure 3.6

We have all flags defined for the device. Now we only need to design the HMI screen graphic i.e. a pictorial view of the S/N and E/W streets, traffic lights, the START, STOP, Emergency pushbuttons, current time, date display PCBs, and speedometer PCB (with its two red and green LED indicators). Figure 3.7 displays the design of our traffic light's system screen view on the device.

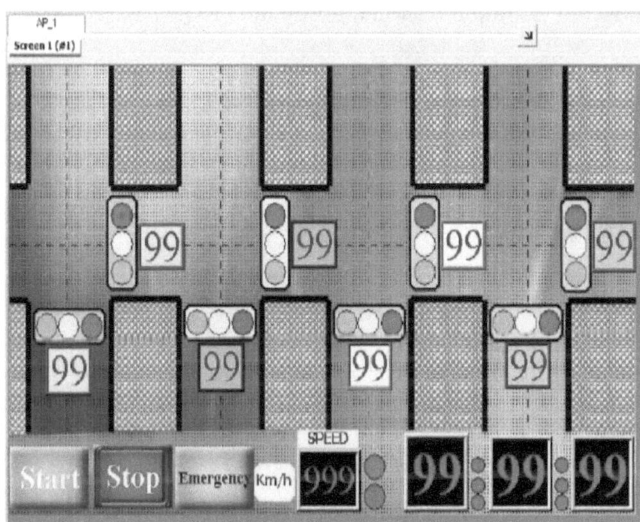

Figure 3.7

Figure 3.7 is the screen view of the traffic light system that is designed to show all the information we can see on the LED display PCBs. Designing such a screen using a state of art software that are developed by the maker of HMI systems is very simple.
If you download part of the control program till end of chapter 2 into PLC's memory and RUN, you are going to see all the good things happening with the system. It starts controlling the 4 intersection's traffic lights in harmony.

All signal lights (red, green, and yellow) turn on and off with all countdown timers busy showing all the timings related to the each intersection.
You also can see the current **time** and **date** of the day information displayed on the 7- segment display PCB.

Screen design of HMI device

Application of a HMI display is very simple. From figure 3.7, notice that I have used **PM Designer** software to draw all the street's images plus all the traffic lights, speedometer, time and date display indicators, the **START**, **STOP**, and the **Emergency** pushbuttons.
To draw traffic light symbols, from PM Designer > Draw > Circle, double left click on **circle** image to get to **Circle** dialog box to do the setting. The **red traffic light** is located on the first intersection on **S/M** Main Street. That particular light is connected to output terminal port number **Q1.7** based on figure 2.2 (r_111).

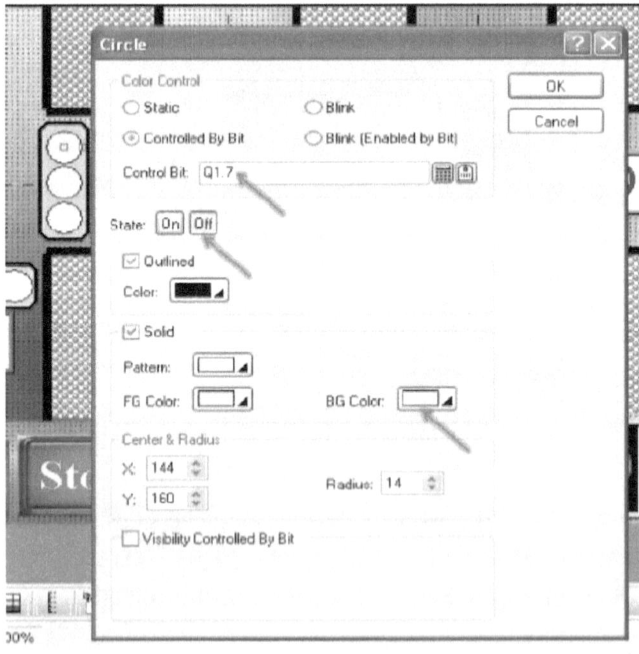

Figure 3.8

In figure 3.8, type **Q1.7** as address of the red traffic light for **Control Bit**. We want the light to look in **white** when it is **off** and **red** when it is **on**. See figure 3.9. These steps should be repeated for all other traffic lights as well.

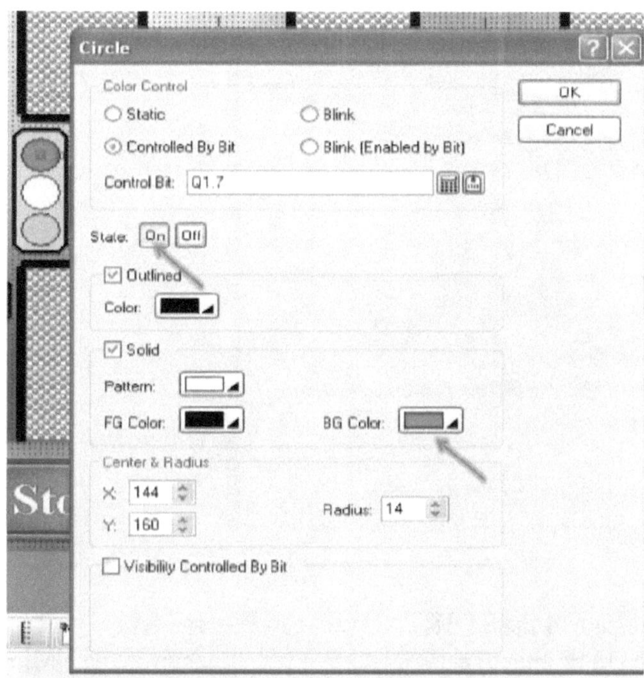

Figure 3.9 – setting **red** traffic light's color when it is turned on.

From PM Designer > Object > **Numeric Display**. And after dragging the display object into the graphic area, double left click on it to do the setting when the **Numeric Display** dialog box appears.

From network 85, you see when M0.1 + M0.2 = 1, G1 = 1, content of (MW20) is going to be written in (MW52). To show the content of MW52 on the related HMI display, do the setting as done on figure 3.10. The same steps should be taken to show content of other memory words on all other 7-segment displays (7 other countdown timers).

Actually, the MOVE instruction is used on Networks 85 to 100 (at phase 10) transfers a content of a previous memory word to a new one which would be displayed at the HMI's Numeric Display object for each different intersection.

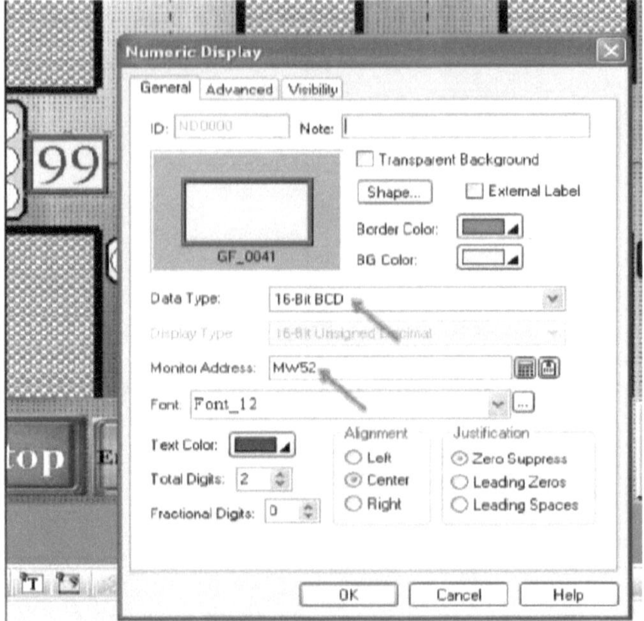

Figure 3.10

Now it is time to take care of the **START**, **STOP**, and **Emergency** pushbuttons. We need to be able to activate these pushbuttons from the HMI device too.

Figure 3.10A displays the assignment of the flags to each of pushbuttons used in the HMI program.

Network #	Pushbutton function	Flag #
173	Emergency	M132.3
176	START	M132.1
177	STOP	M132.2

Figure 3.10A

Screen 1 (#1)
Bit Button
General Label Advanced Visibility
ID: BB0000   Note:
Picture Shape   External Label
Shape...   VFTA: Sunken
State: On Off   Test VFTA
Border Color: FG Color:
Pattern: BG Color:
NO_BDR
Operation
Set ON   Set OFF   Set ON Pulse   Set OFF Pulse
Momentary ON   Momentary OFF   Invert
Address Type: Bit
Write Address: M130 1
Pulse Width: 0.05 second
Monitor   Monitor Address identical to Write Address
Monitor Address: M130 6
ON Macro
OK   Cancel   Help
100 x 68   100

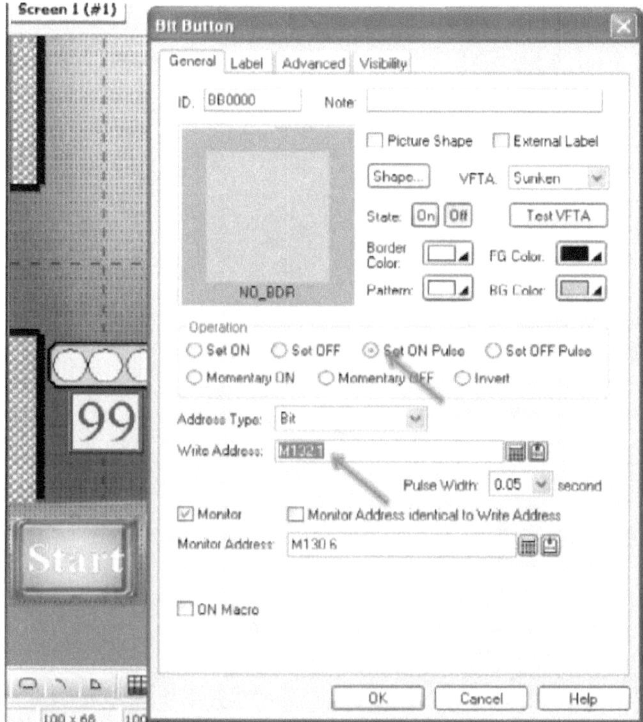

Figure 3.11

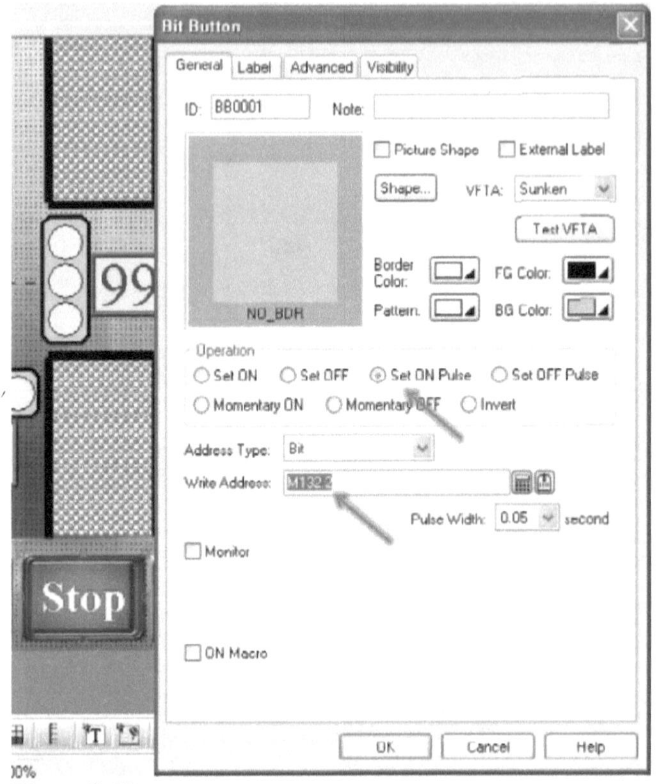

Figure 3.12

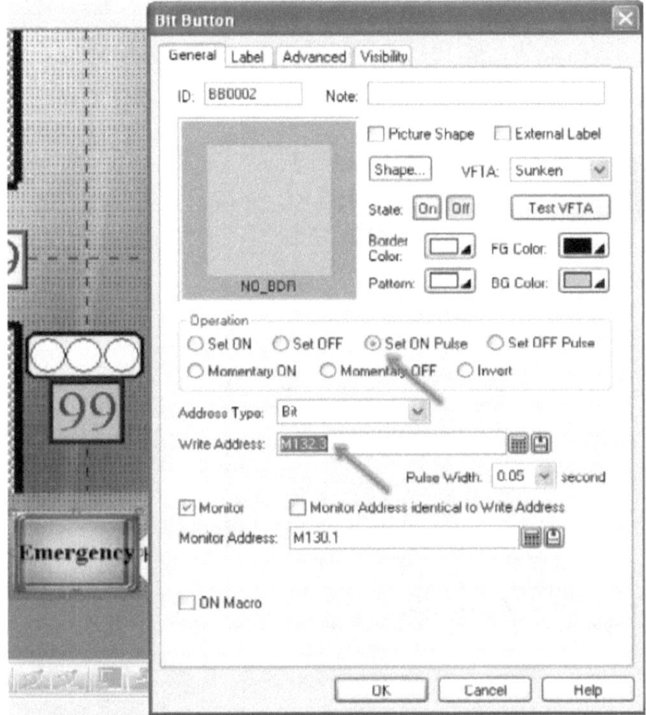

Figure 3.13

To display value of a speed calculated while a car passing from S1 to S2, we can again use another **Numeric Display** Object and define content of MW90 as a **Monitor Address** of the HMI device. Actually, the content of MW90 is a number which is calculated based on the time it takes for the car to travel 100 meters. So content of (MW90) is a number which is = XX Km/h of the car which would be displayed on the **SPEED** Numeric Display Object which represents our **7-segment display PCB** installed on the hardware panel.

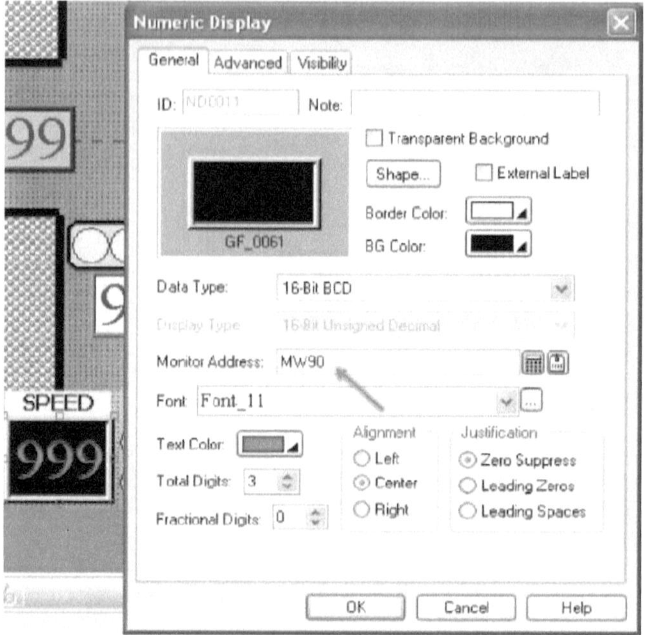

Figure 3.14

Remember that if a car passes with speed >= 30 Km/h, the green LED light is supposed to be turned on otherwise, red one is turned on. Also the value of calculated speed would be displayed.

Figures 3.15 and 16 show how the related setting for red LED is done.

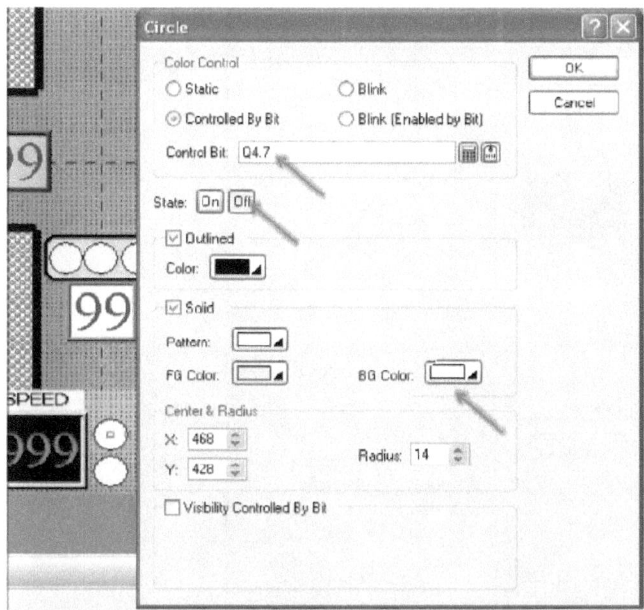

Figure 3.15

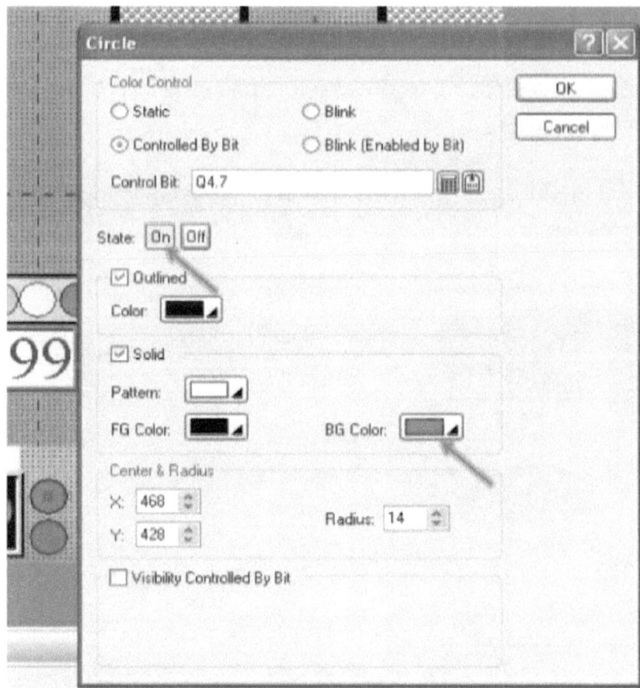

Figure 3.16

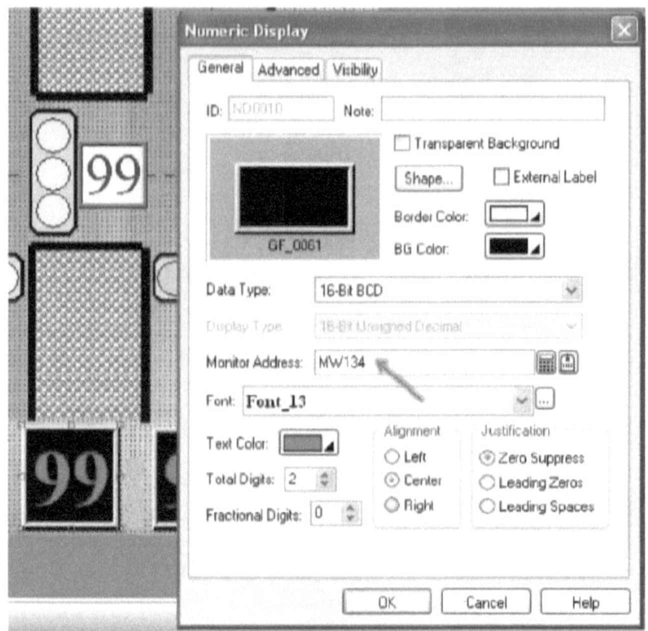

Figure 3.17

Assuming on this phase, HMI's software is complete, now it is time to download the program into its memory as it was explained on phase 18 explained earlier (see figure 3.18).

Figure 3.19 displays the final look of the HMI screen when the development of the program is completed.

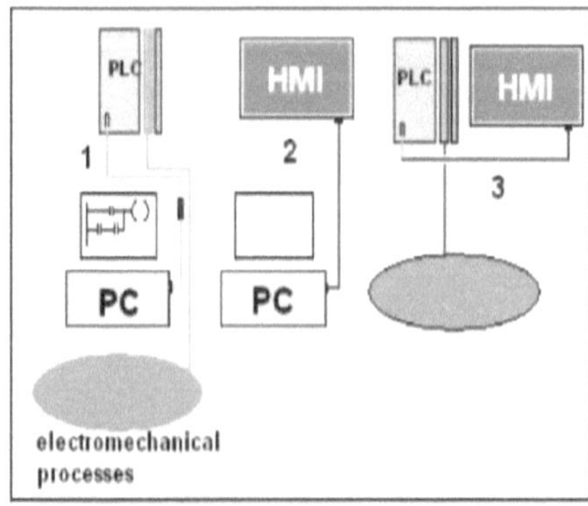

Figure 3.18

Figure 3.19

Application of the WINCC Explorer as an HMI task

In this chapter, we are going to develop the SIMATIC WINCC Explorer HMI program to see also all good things happening on PC's screen i.e. depression of pushbuttons, calculation speed of a traveling car from S1 to S2

I am going to explain briefly how a WINCC Explorer HMI application program can be developed for the project. BUT if you are stuck up anywhere while designing the screen graphic on your own, or having difficulty to understand what am I talking about, you may install WINCC Explorer program on your system and take a look at the settings by checking out the final program I developed from the following directory if you received the CD ROM when you purchased the text.
The name of the file developed for this section is Stop_light.mcp.

<u>**Phase 19 - Developing an HMI application with SIMATIC WINCC Explorer**</u>

The PLC's hardware configuration should be changed to allow PC running WINCC Explorer program to communicate with the PLC through the MPI port.
From the SIMATIC Manager double click on > Hardware. See <u>figure 4.1</u>.

Figure 4.1

Double Click on **CPU 315-2 DP** ... See <u>figure 4.2</u>.

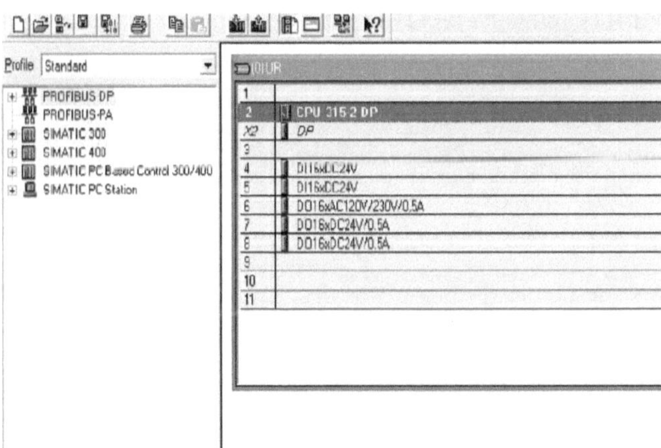

Figure 4.2

Click on **Properties** icon (see <u>figure 4.3</u>).

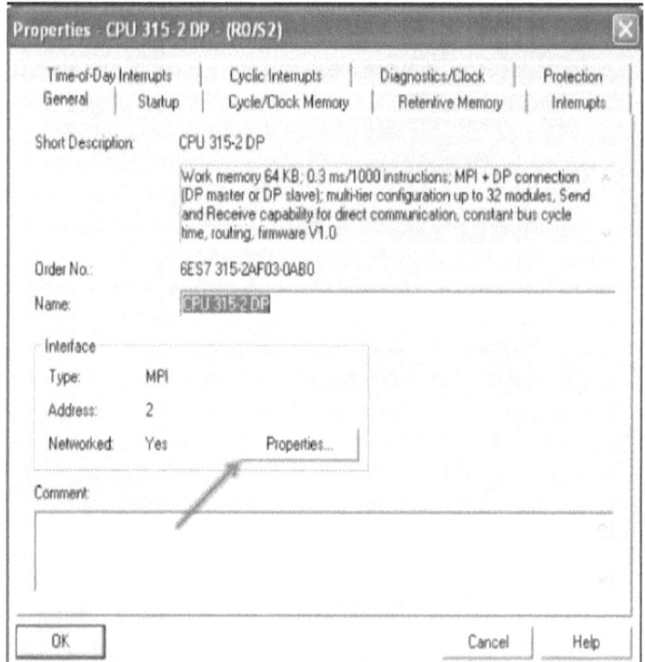

Figure 4.3

In figure 4.4 select **MPI(1)** and click on **OK** icon.

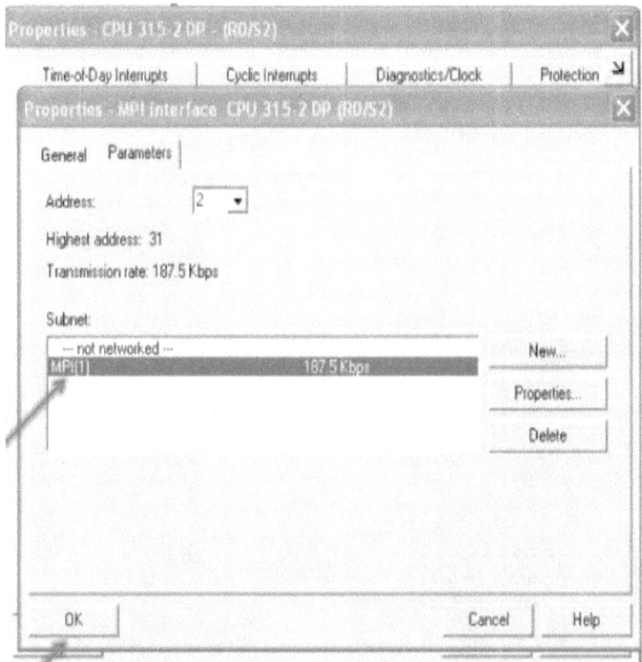

Figure 4.4

And at the end, **File > Save and Compile**.

The SIMATIC WinCC Explorer software can also be used to develop an HMI application similar to the one we already developed for the HMI device developed by PanelMaster Company.
The WINCC Explorer program functions similar to the HMI device which was used already to display timing information speed of the cars passing S1 and S2 IR sensors, depression of pushbuttons etc. Hence, by adding some more coding to our previously developed control program, and also developing a new WINCC Explorer based application program, we can get to see all the timing data on our PC's monitor as well.

Figures 4.5 and 6 display networks we need to add to our PLC control program to change the Countdown timer data from BCD format to an Integer one to make it suitable to be used on our newly developed WinCC HMI monitoring application. Figure 4.6A displays table showing its content in Integer form of Memory Words holding the Countdown timer's data.

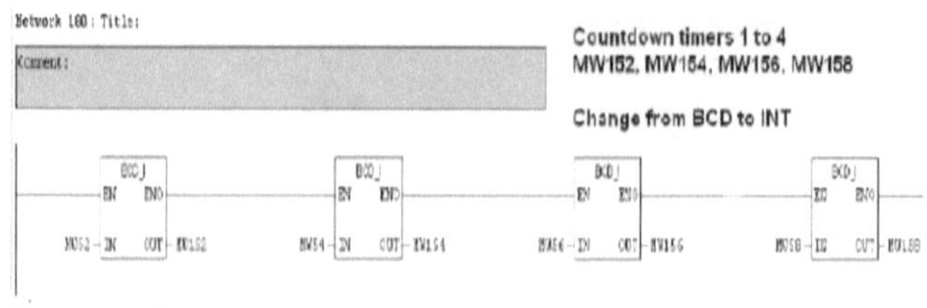

Figure 4.5

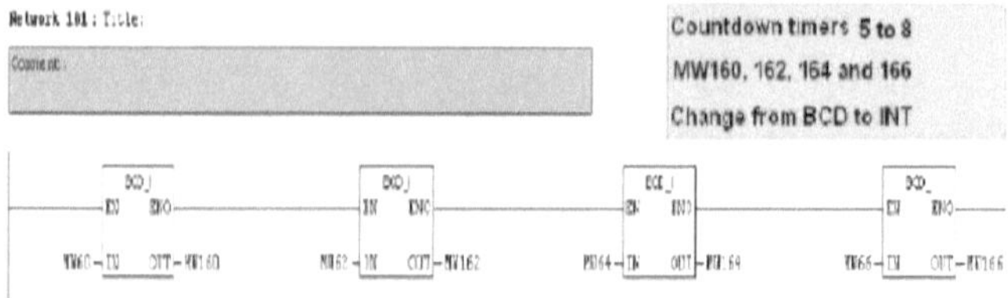

Figure 4.6

Countdown timer #	MWxxx	#	MWxxx
1	152	5	160
2	154	6	162
3	156	7	164
4	158	8	166

Figure 4.6A

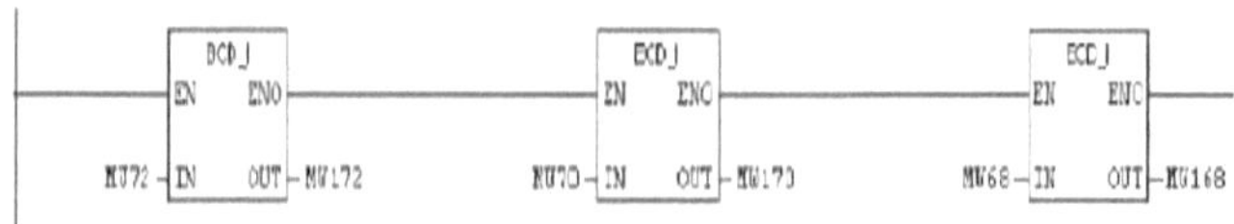

Figure 4.7

Figure 4.7A displays table showing content of the memory words we need to use in our application program (data related to speed: Km/h, time it takes to drive from S1 to S2, year, month, and day).

Remember one other requirement is to calculate the speed of a car passing between S1 and S2 with a distance of 100 meters (on S/N Main Street) and display it on the 9th

7-segment display PCB. Showing the speed value is supposed to last about 20 seconds and then it is going to be replaced with **00**. If speed value is even 1 Km/h greater than 30, the red LED light comes on and 31 number start turning on and off (FFH is going to be sent on the display). Table in <u>figure 4.7A</u> displays the content of flags or MVs used in the control program to hold the related data used by the WinCC application program.

Description	MW xxx	Network #
Speed Km/h	MW190	185
Time between S1/S2	MW180	183
START	M132.1	176
STOP	M132.2	177
Emergency	M132.3	173
year	Mw168	182
Month	MW172	182
Day	MW170	182

Figure 4.7A

Network 183 : Title:

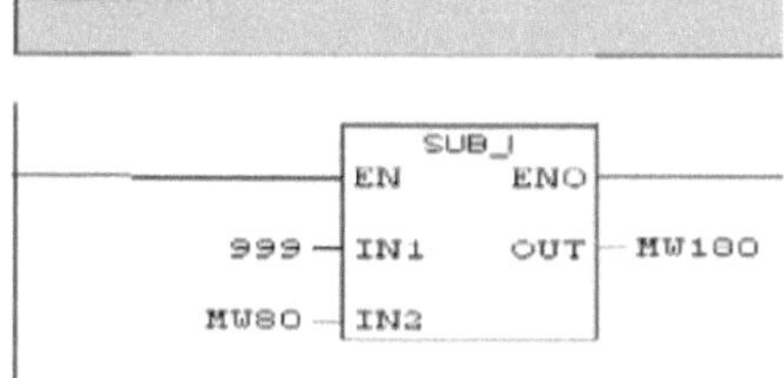

Figure 4.8

Network 184 : Title:

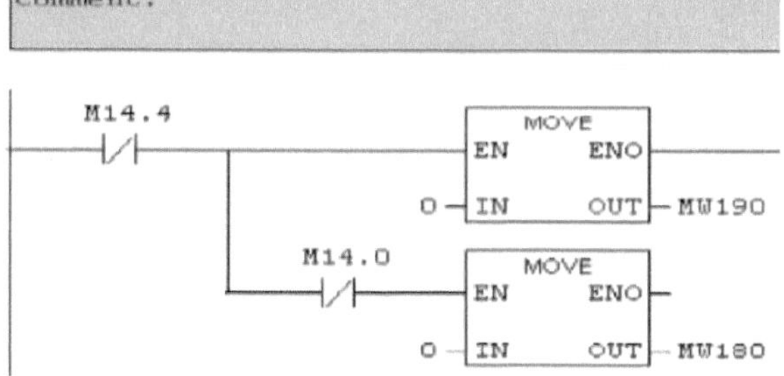

Figure 4.9

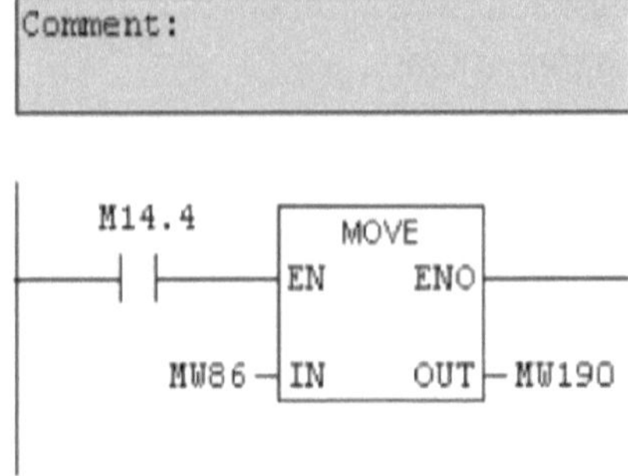

Figure 4.10

Now that all codes related to the PLC control program is generated, it is time to fix codes or setting parameters to display all data related to the countdown timers, speed etc on our PC's monitor by application of the WinCC software. Certain settings must be done before running the developed WinCC software.

To start defining new tags, do the following steps:

1- In **WinCC Explorer**, click on **File** and when WinCC Explorer dialog box appears, check **Single-User Project** then click **OK**

Figure 4.10A

2- On **Create a new project** dialog box, type **Project Name** and click **Create**.

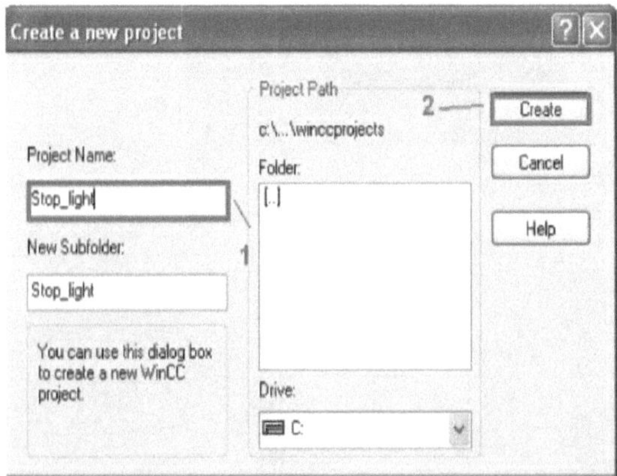

Figure 4.10B

3- In the WinCC Explorer, click right Tag Management > click Add New Driver.

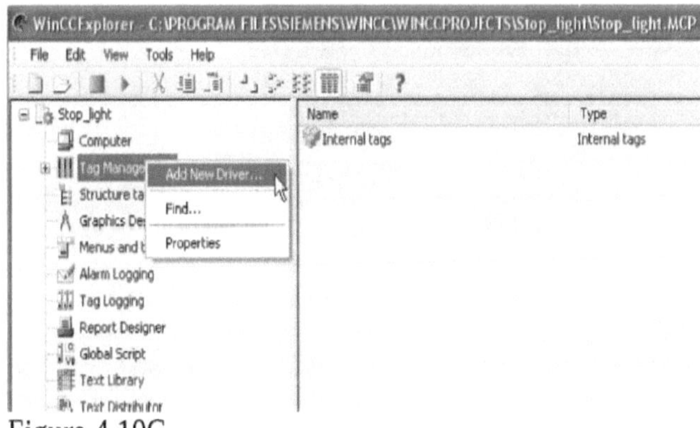

Figure 4.10C

4- In Add new driver dialog box, click the SIMATIC S7 Protocol Suite.chn >

Click **Open**.

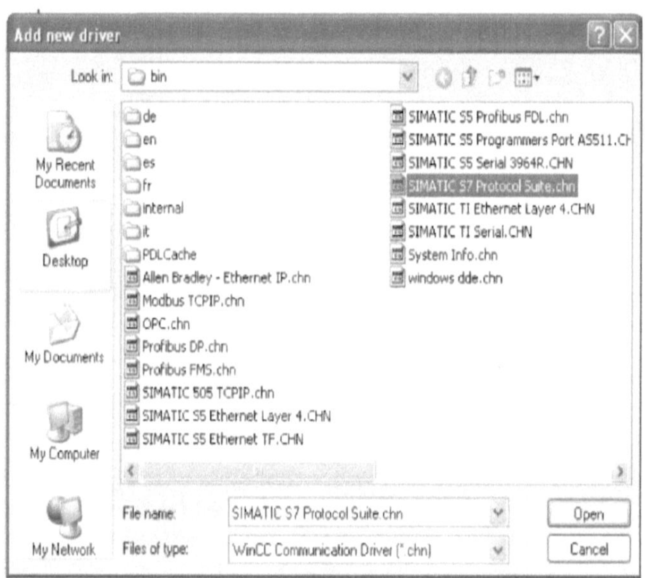

Figure 4.10D

5- Right click on **MPI** and click on **New Driver Connection…**

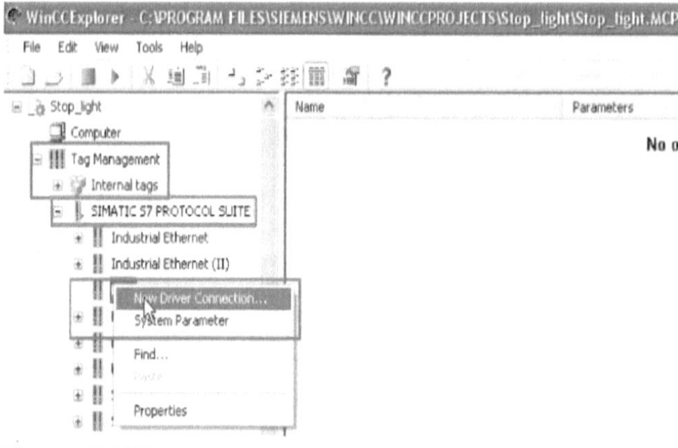

Figure 4.10E

6- When **Connection properties** dialog box appears, if you do not want to change **New Driver Connection**… to anything else, click **OK**. See next figure 4.10F.

Figure 4.10F

7- Right click on NewConnection, click on **New Tag.**.

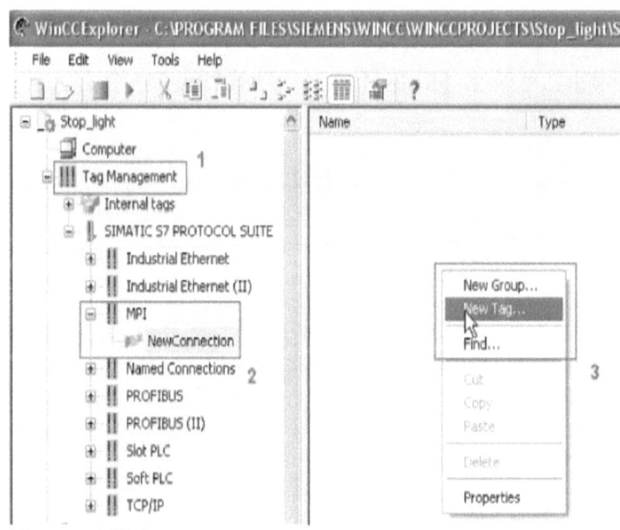

Figure 4.10G

8- In **Tag properties** dialog box, type tag name **(R_111)** and choose Data Type: **(Binary tag)**. Click on Select, to define **Address**.

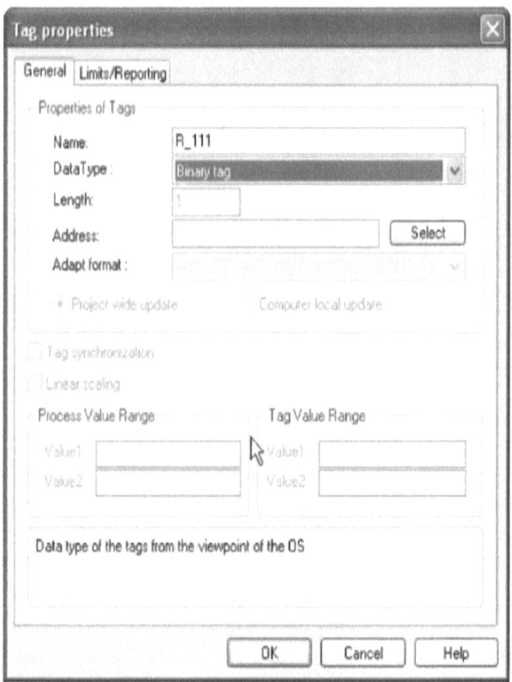

Figure 4.10H

9- At Address properties dialog box, define **Data**, and **Address**, **Q, Bit**, click **OK**.

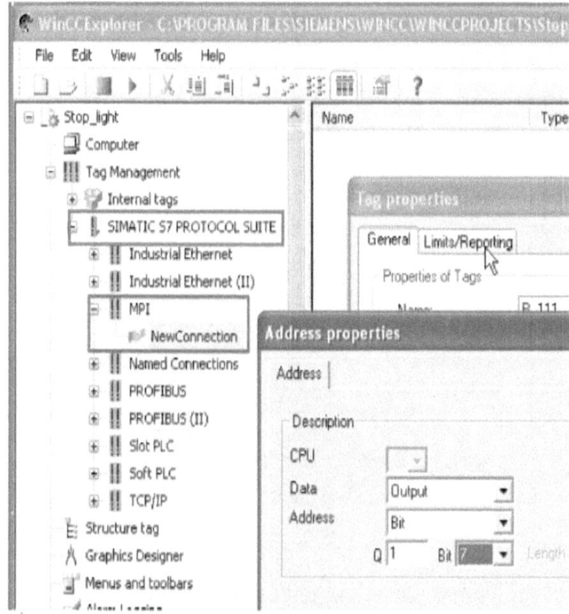

Figure 4.10J

10 – After **OK** is clicked, **R_111** tag is defined (figure 4.10K) with its properties as shown. To create any other new tag, just repeat the same steps from 7 to 10.

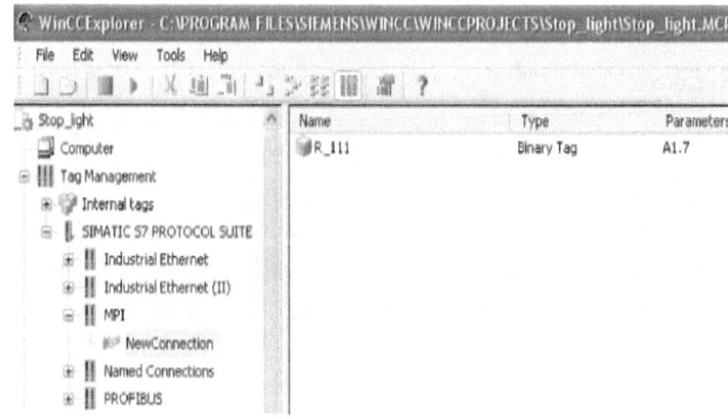

Figure 4.10K

Figure 4.10 displays list of all tags defined for this project.

Name	Type	Parameters
R_111	Binary Tag	A1.7
Y_112	Binary Tag	A2.0
G_113	Binary Tag	A2.1
R_121	Binary Tag	A2.2
Y_122	Binary Tag	A2.3
G_123	Binary Tag	A2.4
R_211	Binary Tag	A2.5
Y_212	Binary Tag	A2.6
G_213	Binary Tag	A2.7
R_221	Binary Tag	A3.0
Y_222	Binary Tag	A3.1
G_223	Binary Tag	A3.2
R_311	Binary Tag	A3.3
Y_312	Binary Tag	A3.4
G_313	Binary Tag	A3.5
R_321	Binary Tag	A3.6
Y_322	Binary Tag	A3.7
G_323	Binary Tag	A4.0
R_411	Binary Tag	A4.1
Y_412	Binary Tag	A4.2
G_413	Binary Tag	A4.3
R_421	Binary Tag	A4.4
Y_422	Binary Tag	A4.5
G_423	Binary Tag	A4.6

Figure 4.10

Figure 4.11 displays name of all Tags related to the countdown timers and the different intersections.

Name	Type	Parameters
t11	Unsigned 16-bit value	MW152
t12	Unsigned 16-bit value	MW154
t21	Unsigned 16-bit value	MW156
t22	Unsigned 16-bit value	MW158
t31	Unsigned 16-bit value	MW160
t32	Unsigned 16-bit value	MW162
t41	Unsigned 16-bit value	MW164
t42	Unsigned 16-bit value	MW166

Figure 4.11

Figure 4.12 displays assignment of all **Tags** related to speed, related red and green **LED** indicators, **Day, Month, Year, START, STOP, Emergency** etc.

speed	Unsigned 16-bit value	MW190
Year	Unsigned 16-bit value	MW168
Day	Unsigned 16-bit value	MW170
month	Unsigned 16-bit value	MW172
start	Binary Tag	M132.1
stop	Binary Tag	M132.2
eme	Binary Tag	M132.3
LED_red	Binary Tag	A4.7
LED_green	Binary Tag	A5.0
Time	Unsigned 16-bit value	MW180

Figure 4.12

Now it is time to design the graphic of the screen in which we want our traffic light hardware system to be shown on our PC's monitor.

Figure 4.13 displays a simple graphic designed to simulate the traffic light hardware system.

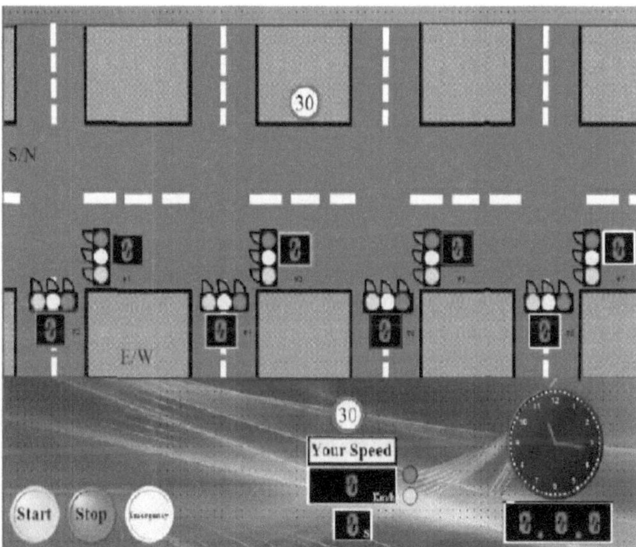

Figure 4.13

Figure 4.14

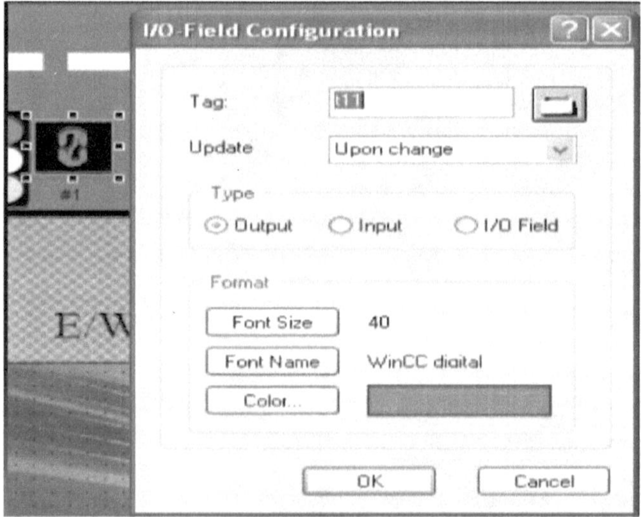

Figure 4.14

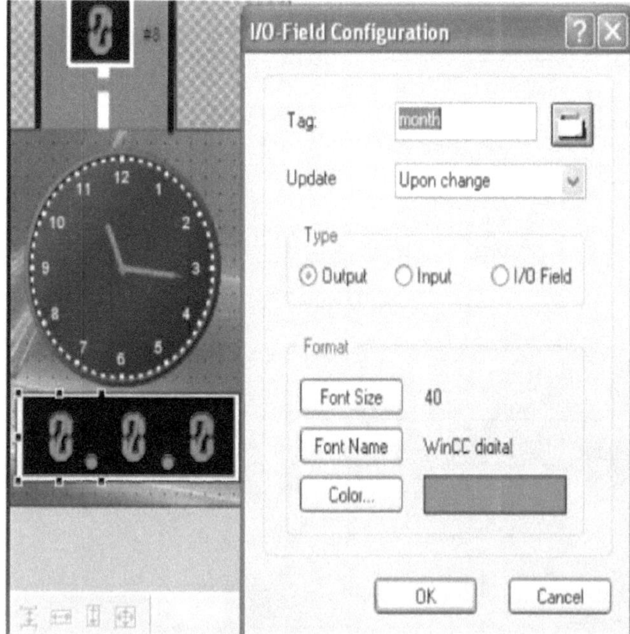

Figure 4.15

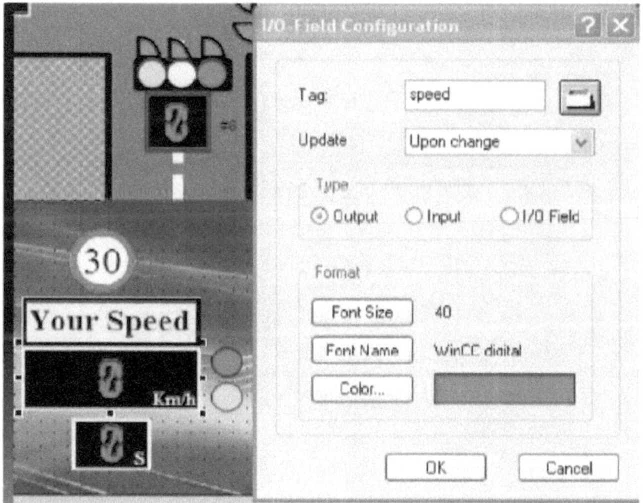

Figure 4.16

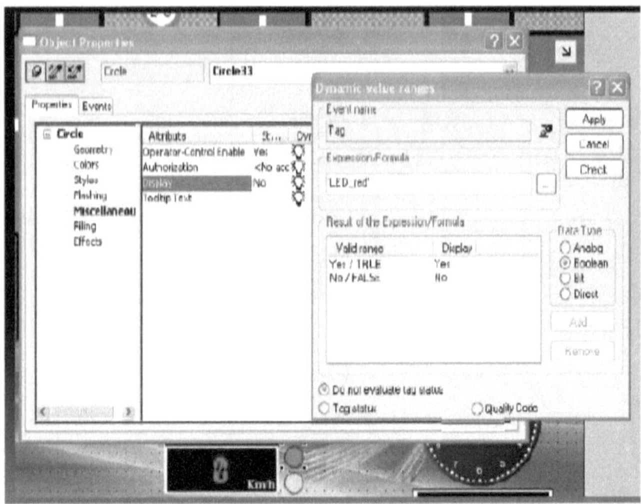

Figure 4.17

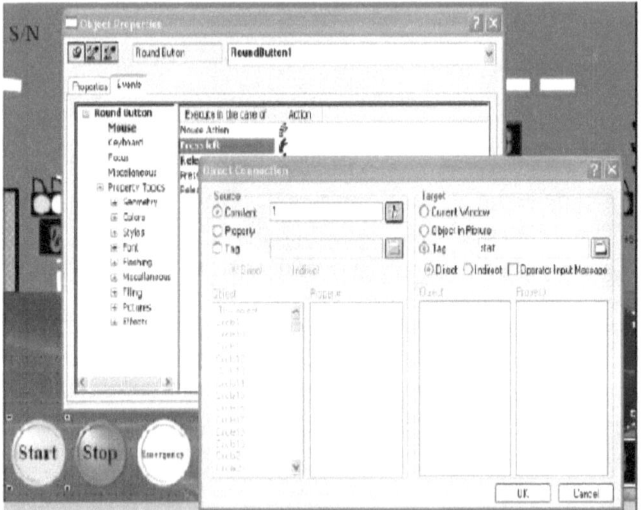

Figure 4.18

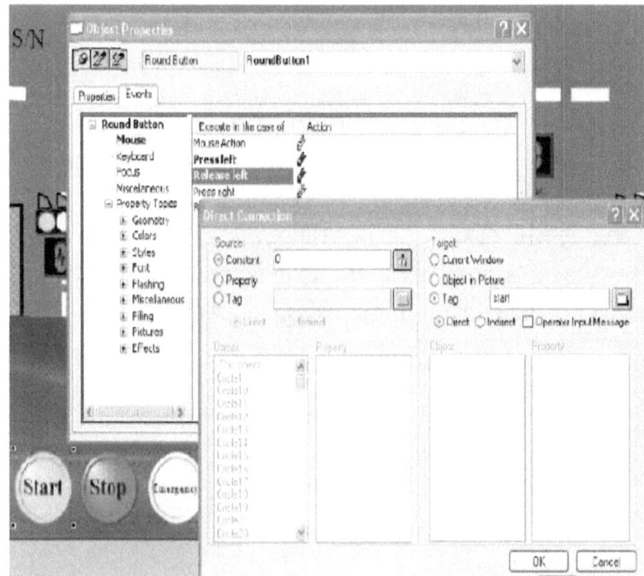

Figure 4.19

This section of this manual is reserved for you. I would be very happy if
you would like write to me with any questions (hardware or software wise)
that you have related to this well designed industrial project.
I shall answer you question ASAP and also, will include your question in this
section of the manual, as I believe that it might be also somebody else's
question, as well.

Thanks for your comments and concerns regarding this design!

Now you have the information you need to become a successful PLC programmer. This is a moment for you to take the information you've been given and not be afraid to put it to use.

The primary thing that keeps people from becoming successful is fear. If you can overcome this fear then you will succeed and prosper.

All you need to do at this point is to stop putting things off and just do it. Start working and you will find it becomes easier as you go along.

Once you accomplish any new skill or task, you will realize just how easy it is. That is when you want to kick yourself for not starting sooner.

So don't hesitate. Start right now. Start today. Break that cycle of doubt and enjoy your newfound skills and succeed in exploiting them.

To your success and happiness,

Seyedreza

Hardware, software & more

We would like to introduce our Teach Yourself Consumer PLC books as an excellent resource for readers who want to learn more about PLCs. So far, we have launched this groundbreaking new series with five exciting titles:

1- Programming with SIMATIC S7 300/400 Programming Controllers
2- Programming with SIMATIC S7-200 PLCs using STEP 7-Micro/win software
3- Basics of Programmable Logic Controllers (PLC)
4- Programming Alley Bradley series of PLCs using RSLogix 500 software compiler

Design and Implementation of typical end year college projects using PLC or a typical microcontroller with youtube.com site links to watch the projects in action:

1- Programming a 4 floor elevator with SIMATIC STEP 7 & LOGO! SoftComfort software
http://youtu.be/dqkeHD5WNcc
2- Programming SIEMENS LOGO! PLC as a dyeing machine temperature controller
http://youtu.be/mJfT4z1oCeo
3-Programming a Traffic Light Controller system using SIMATIC S7-300 & HMI display
http://youtu.be/0ADfFPOzIUE
4- Programming a PLC based car Parking System with SIMATIC STEP 7 and S7-300 PLC
http://youtu.be/51oqLRxXcHk
5- Programming a Traffic Light Control system using an AVR ATmega8 microcontroller
http://youtu.be/q0eWsX2kMeU
6- Temperature controller and monitoring with a microcontroller
http://youtu.be/mIyixludYaE

7- Microcontroller based 4 Floor Elevator System
http://youtu.be/UQbGcntcVP4
8- How to control a Stepper motor with an ATmega8 microcontroller
http://youtu.be/XBWCsl512Lc
9- MATLAB/PC and Microcontroller based advanced Line following Robot
http://youtu.be/2Y_qSzKL_2Y
10- Analog LED clock with chime, alarm, calendar and temperature display
http://youtu.be/Mi5zAdctp4g
11- Design and implementation of a Microcontroller-Based Car Parking System
http://youtu.be/7aIp2iCC_mU
12- 2 Zones microcontroller based weekly digital timer
http://youtu.be/xbIEKI3qqNM
13- Implementing an arduino based temperature controller with PID algorithm http://youtu.be/i2nxqLs9wBg
14- Implementing a PLC-based temperature controller with PID algorithm
http://youtu.be/yhSyKObMlgA

www.ingramcontent.com/pod-product-compliance
Lightning Source LLC
Chambersburg PA
CBHW021005160726
47994CB00006B/2377